J. Holland

A preliminary revision and synonymic catalogue of the

Hesperiidae of Africa and the adjacent islands

With descriptions of some apparently new species

J. Holland

A preliminary revision and synonymic catalogue of the Hesperiidae of Africa and the adjacent islands
With descriptions of some apparently new species

ISBN/EAN: 9783741114021

Manufactured in Europe, USA, Canada, Australia, Japa

Cover: Foto ©Klaus-Uwe Gerhardt /pixelio.de

Manufactured and distributed by brebook publishing software
(www.brebook.com)

J. Holland

A preliminary revision and synonymic catalogue of the Hesperiidae of Africa and the adjacent islands

The following papers were read :—

1. A Preliminary Revision and Synonymic Catalogue of the *Hesperiidæ* of Africa and the adjacent Islands, with Descriptions of some apparently new Species. By W. J. HOLLAND, Ph.D., F.Z.S., F.E.S., &c., Chancellor of the Western University of Pennsylvania.

[Received November 6, 1895.]

(Plates I.-V.)

Having been for a number of years past engaged in the diligent study of the Lepidoptera of Tropical Western Africa, and having been compelled in the prosecution of these studies to acquaint myself with the entire literature of the subject, it has occurred to me that it might facilitate the labours of others, who may be tempted to embark upon the same line of investigation, or who may already be involved in the tangled mazes of the subject, if I should at least attempt to bring together into one paper the scattered references to the various species.   I have therefore begun a synonymic catalogue of the Diurnal Lepidoptera of the African Continent and the adjacent Islands, but am led by the advice of trusted friends to anticipate the publication of the more extended catalogue by the following paper, in which is contained a list of a very difficult group of Butterflies included in the fauna.   I am led the more readily to take this step in view of the results of the recent labours of Lieut. E. Y. Watson, who, in a paper recently published in the Proceedings of the Zoological Society upon the Classification of the *Hesperiidæ* (P. Z. S. 1893, p. 3), has laid solid foundations for the prosecution of systematic researches in the future.   I have in the main followed the classification which he has suggested in his valuable paper, which, while confessedly incomplete, and leaving some things to be desired, is, nevertheless, one of the most notable contributions to the literature of the subject which has recently appeared.   Based, as it is, upon an accurate and painstaking examination of the anatomical details and structural peculiarities of the various species represented in the collections of the British Museum and the magnificent collection of Messrs. Godman and Salvin, it may in the main be accepted as free from the blemishes which characterize much of the work done in this group by authors, who have relied almost wholly upon superficial resemblances.   In the few cases in which I have departed from the classification of Lieut. Watson, it has been because I have been able to make more careful anatomical investigations than it was possible for him to do with the material at his command.   A private collector may do as he will with his own, and may bleach and dissect specimens, when it would be little less than a crime for the authorities of a Museum like that at South Kensington to allow such treatment to be bestowed upon the

West, Newman chromo lith

African Hesperidæ.

African Hesperidæ.

African Hesperiidæ.

African Hesperiidæ.

African Hesperidæ.

precious types of Hewitson and other great naturalists, who have placed their collections in the care of the institution.

In following up my labours I have been greatly aided by the possession of a large mass of well determined Indian material, which I have been accumulating for many years past, and particularly by the possession of the Knyvett collection, for which I am indebted to the generous kindness of Mr. Andrew Carnegie, my distinguished fellow-townsman, whose interest in all things relating to the advancement of science is well known. I have derived much assistance from the collections which I have received from Mr. William Doherty, the well-known naturalist explorer of the far East, and from the collections for which I am indebted to Mr. L. de Nicéville, of Calcutta, whose great work upon the Lepidoptera of India is a monument to his painstaking diligence and scientific acumen. I am no less indebted to Mr. Roland Trimen, the late learned Curator of the South-African Museum at Capetown, whose labours upon the fauna of extra-tropical Africa are classic, and who with the most engaging kindness has presented me with authentically determined specimens of most of the species named by him. It is much to be wished that all authors might acquire those habits of exact observation and clear description which are possessed by this Nestor among lepidopterists, whose diagnoses of the various species contained in his last work upon the Butterflies of South Africa are so exact as almost to make the work of pictorial representation superfluous. I am under very special obligations to the authorities of the British Natural History Museum not only for permission to freely study the collections in their possession, but for permission to have drawings made of the hitherto unpublished types of the late Mr. Hewitson and of Dr. Butler. I have to thank Dr. Karsch of the Berlin Museum, and Dr. Rogenhofer of the Imperial Museum at Vienna, for similar kindnesses. From Mons. Mabille of Paris I have received most distinguished courtesies, and I am indebted to him for the opportunity to examine personally the types of many of his recently described species, and for the use of a number of copies of the unpublished figures of Ploetz. Ploetz made no collection of specimens during his lifetime, but contented himself with making drawings, not always very accurate, of the species which he described in the collection of others, or which he found figured in various works. These figures are in many cases our only safe clue to a knowledge of the species he named, for his descriptions are in many instances very unsatisfactory. I cannot fail in this connection to express my indebtedness to Lieut. Watson, who compared many of the species in my collection with the types in the British Museum, and assigned them to the respective genera to which they belong in his classification, and to Dr. Butler and Mr. Herbert Druce for their generous assistance at all times freely given. Among American entomologists, I am especially indebted to Dr. S. H. Scudder of Cambridge, who, upon the occasion of his last visit to Europe, did me the great favour of comparing a series of drawings of the species

1*

in my collection with the types in the Berlin Museum and in the Museums of Paris and London. But great as is the debt of gratitude I owe to these valued friends and colabourers, it is even exceeded by my obligations to Dr. Otto Staudinger of Dresden, who entrusted to the ocean all the types of African *Hesperiidæ* and all the unnamed material in his vast collection, and freely sent them to me for purposes of study and comparison. For this act of great generosity I cannot sufficiently thank him.

In submitting the following pages to the attentive consideration of specialists, it is with a sense of the manifold defects which must in the lapse of time be found to be contained therein. With the exercise of the utmost care, and with all the help of the learned, errors are unavoidable. In all cases where doubt attaches in my mind to a generic reference, it is indicated. Absolute certainty in this respect is not easily attained in some cases. While two-thirds of the species accredited to the African fauna are represented in my own collection, in some cases by enormously large series of specimens, and I have seen in nature probably four-fifths of the species of the *Hesperiidæ* which have been described as coming from Africa, nevertheless in not a few cases I have been compelled to rely wholly upon illustrations and the suggestions of resemblance made by authors for an approximate location of the species. Yet, in spite of the defects which must of necessity exist in this work, I venture to express the confident belief that it will be found to mark a distinct advance in our knowledge of the subject.

RHOPALOCERA.

Fam. HESPERIIDÆ.

Subfam. HESPERIINÆ.

SARANGESA, Moore.

(*Hyda*, Mab.; *Eretis*, Mab.; *Sape*, Mab.)

The differences of a structural character between the species assigned to the genus *Eretis*, Mab., and *Sarangesa*, Moore, are so slight as in my estimation not to justify a separation, except subgenerically. The principle difference is in the waved outline, of the secondaries and the relatively longer fringes in the form *Eretis*.

* ERETIS, Mab.

1. S. DJÆLÆLÆ, Wallgr.

- *Pterygospidea djælælæ*, Wallgr. K. S. Vet.-Akad. Handl. 1857; Lep. Rhop. Caffr. p. 54, no. 5.

Nisoniades umbra, Trim. Trans. Ent. Soc. Lond. (3) vol. i. p. 289 (1862).

Nison. djælælæ, Trim. Rhop. Afr. Austr. vol. ii. p. 311, no. 204 (1866).

Pterygos. djælælæ, S. Afr. Butt. vol. iii. p. 254, pl. xii. fig. 7, ♀ (1889).

Hab. S. Africa.

Lieut. Watson, P. Z. S. 1893, p. 48, calls attention to the fact that the species in the British Museum which has been identified by Mr. Butler from various localities in Northern and Eastern Africa as *S. djælælæ*, Wallgr., is not that species, and is apparently unnamed. This form, which is common in Abyssinia and elsewhere, is more closely related to *S. motozi*, Wallgr., and falls into the subgenus *Sape* of Mabille. Mons. Mabille, I discover, has labelled it as *S. nerva*, Fabr., in the collection of Dr. Staudinger, and so also has labelled it for me. It certainly is not the insect described under this name by Fabricius, and I have therefore ventured elsewhere to name and describe it (vide *S. eliminata*, Holl., p. 9).

2. S. LUGENS, Rogenhfr. (Plate II. fig. 10.)

♀ (♂ sec. Rghfr., sed non sec. Rebel). *Pterygospidea (Tagiades,* Ploetz) *lugens*, Rogenhfr. Ann. Hofmus. Wien, vol. vi. p. 46 (1891).

♂ . *Pterygospidea morosa*, Rogenbfr. Ann. Hofmus. Wien, vol. vi. p. 463 (1891).

Hab. Marangu, Tropical Africa (*Von Hoehnel*).

I am under profound obligations to Dr. Rogenhofer, of the Imperial Museum in Vienna, and to Dr. Rebel, his assistant, for having kindly furnished me with most carefully executed drawings of the two forms characterized as above by Dr. Rogenhofer. Dr. Rebel writes me as follows :—" I have taken occasion to critically examine the two unique types of *P. lugens*, Rghfr., and *P. morosa*, Rghfr., and have positively ascertained that both names apply to one species. The name *lugens*, Rghfr., must stand, inasmuch as it is the first in the order of publication. Rogenhofer is in error in regarding the type of *lugens* as a male ; it is most positively a female. The name *morosa* must therefore sink as a synonym (= ♂ of *lugens*)."

3. S. MELANIA, Mab.

Eretis melania, Mab. C. R. Soc. Ent. Belg. 1891, p. lxxi ; Watson, P. Z. S. 1893, p. 48 ; Karsch, Berl. Ent. Zeit. Bd. xxxviii. p. 264, ♀ ? (1893).

Hab. Gaboon ; Togoland ?

Dr. Karsch refers a female before him with doubt to the species described by Mabille. In the vast series of specimens which I have received from Gaboon, I have never found one which tallies exactly with the type or description of Mons. Mabille. I thought that the following species might be the same, but having compared my type with the original type of *E. melania* in the collection of Dr. Staudinger, I am quite sure of the distinctness of the two species. *S. melania* may be readily distinguished from *S. perpaupera*, which it closely resembles at first sight, by the fact that the fringes of the primaries, which are fuscous, are checkered with black at the ends of the nervules, and are conspicuously white at the apex and at the inner angle. The specimen in Dr. Staudinger's collection is labelled "*melanina*" in the handwriting of Mons. Mabille. The published name is *melania*, and this of course stands.

4. S. PERPAUPERA, Holl.

Sarangesa perpaupera, Holl. Ann. & Mag. Nat. Hist. (6) vol. x. p. 288 (1892); Ent. News, Jan. 1894, pl. i. fig. 1, ♂.

Hab. Upper Valley of the Ogové River (*Good*); Angola (*Staudinger*).

5. S. EXPROMPTA, Holl.

Sarangesa exprompta, Holl. Ent. News, Jan. 1894, p. 26, pl. i. fig. 3, ♂.

Hab. Accra.

The type was purchased from Doncaster with a lot of other African material. Whether the locality label attached to the specimen is correct I cannot be positively certain, as some of the things bought at the time were plainly not from the localities indicated upon the labels.

6. S. ASTRIGERA, Butl. (Plate II. fig. 8.)

S. astrigera, Butl. P. Z. S. 1893, p. 669.

Hab. Zomba, British Central Africa.

I only know this species by the description given by the author, and the figure prepared by Mr. Horace Knight, which is reproduced upon the plate. I place it in this section of the genus with much doubt, but it plainly belongs here, rather than elsewhere.

** HYDA, Mab.

7. S. GRISEA, Hew.

Pterygospidea grisea, Hew. Ann. & Mag. Nat. Hist. (5) vol. i. p. 344 (1878).

Hyda micacea, Mab. Bull. Soc. Ent. France, (6) vol. ix. p. clxvii (1889); Novit. Lepidopt. p. 93, pl. xiii. fig. 3 (1893).

Hab. Gaboon, Liberia.

Weymer in Stübel's 'Reise,' p. 126, pl. iv. fig. 5, describes and illustrates a species from Ecuador as *Hesperia micacea*. Inasmuch as Mabille's name drops as a synonym in the case of the present species, that of Weymer should be allowed to stand for the species he named.

8. S. TRICERATA, Mab.

Hyda tricerata, Mab. C. R. Soc. Ent. Belg. 1891, p. cvi; Novit. Lepidopt. p. 92, pl. xiii. fig. 2.

Hab. Sierra Leone, Cameroons, Gaboon.

9. S. MAJORELLA, Mab.

Hyda majorella, Mab. C. R. Soc. Ent. Belg. 1891, p. cvii; Novit. Lepidopt. p. 92, pl. xiii. fig. 1 (1893).

Eretis motozi, Wallgr. (?), Karsch, Berl. Ent. Zeit. vol. xxxviii. p. 264, pl. vi. fig. 11 (1893).

Hab. Sierra Leone (*Mabille*); Togoland (*Karsch*).

*** SAPE, Mab.

10. S. LUCIDELLA, Mab.   (Plate II. fig. 22.)

Rape lucidella, Mab. C. R. Soc. Ent. Belg. 1891, pl. lxvii.

The type specimen in the collection of Dr. Staudinger is somewhat worn, but shows that the insect is abundantly distinct from the other species herein enumerated. This is brought out clearly in the figure given.

11. S. MOTOZI, Wallgr.

♀. *Pterygospidea motozi*, Wallgr. K. Sv. Vet.-Akad. Handl. 1857; Lep. Rhop. Caffr. p. 53; ♀, Trim. Rhop. Afr. Austr. vol. ii. p. 313, no. 206, pl. vi. fig. 3; ♂ and ♀, Trim. S. Afr. Butt. vol. iii. p. 356 (1889).

♀. *Nisoniades pato*, Trim. Trans. Ent. Soc. Lond. (3) vol. i. p. 404 (1862).

Hab. South Africa.

12. S. MOTOZIOIDES, Holl.

Sarangesa motozioides, Holl. Ann. & Mag. Nat. Hist. (6) vol. x. p. 288 (1892); Ent. News, Jan. 1894, pl. i. fig. 5, ♀ ; Butl. (?), P. Z. S. 1893, p. 668.

Hab. Transvaal (*in Staudinger's collection*): Gaboon (*Good*).

The male described by me in the 'Annals and Magazine of Natural History,' and subsequently figured in the 'Entomological News,' turns out to be the male of the species described by Mabille as *Pterygospidea bouvieri*, if thorough reliance may be placed upon the identification made in the collection of Dr. Staudinger by Mons. Mabille, the author of the species. So far I have not been able to find in any collection a true male of *S. motozioides*, Holl. The female may be separated at a glance from the female of *S. motozi* by the absence of the conspicuous translucent spot in the cell of the secondaries, which is characteristic of *motozi*, Wallgr., and by the fact that the translucent spots in the primaries are much smaller than in typical *motozi*.

13. S. SYNESTALMENUS, Karsch.

Antigonus synestalmenus, Karsch, Berl. Ent. Zeit. vol. xxxviii. p. 263, pl. vi. fig. 8 (1893).

This species is very closely allied upon the upper surface to

S. bouvieri, Mab., and *S. pertusa*, Mab., but upon the underside reveals great differences.

Hab. Togoland (*Karsch*).

14. S. PERTUSA, Mab.

Sape pertusa, Mab. C. R. Soc. Ent. Belg. 1891, p. lxviii.

Hab. Transvaal.

The type would seem to indicate that this is only a slight variety of *motozi*, Wallgr.

15. S. BOUVIERI, Mab.

Pterygospidea bouvieri, Mab. Bull. Soc. Zool. France, 1877, p. 239.

Sarangesa motozioides, ♂, Holl. Ann. & Mag. Nat. Hist. (6) vol. x. p. 288 (1892); Ent. News, Jan. 1894, pl. i. fig. 4, ♂.

For the determination of this species I am indebted to Dr. Staudinger, who has loaned me a male and female determined for him by the author of the species. By the description originally given by Mons. Mabille, I should not have been able to reach a positive conclusion, as the description seems to be somewhat inadequate.

16. S. THECLA, Ploetz. (Plate V. fig. 14.)

Antigonus thecla, Ploetz, S. E. Z. vol. xl. p. 361 (1879).

Sape semialba, Mab. C. R. Soc. Ent. Belg. 1891, p. lxvii; Karsch, Berl. Ent. Zeit. vol. xxxviii. p. 262.

By comparison of the type of Mons. Mabille with a figure of the type of Ploetz, which is reproduced in the plates accompanying this article, I am able to positively affirm the identity of the two.

Hab. Aburi (*Ploetz*); W. Africa (*Mabille*); Cameroons (*Good*); Togoland (*Karsch*).

17. S. THEOLIDES, sp. nov. (Plate V. fig. 3.)

♂. Antennæ black, slightly lighter on the underside, the upperside of the palpi, thorax, and abdomen is fuscous. The lower side of the palpi is yellowish. The lower side of the thorax and abdomen is pure white. The legs are white, narrowly edged with blackish upon the anterior margins. The ground-colour of the upperside of the primaries and secondaries is fuscous ochraceous. The primaries are heavily bordered with black on the outer margin, and there is a large irregularly quadrate spot of the same colour on the costa near the end of the cell, limited anteriorly by four minute white translucent subapical spots and posteriorly by three like spots, two of them in the cell near its end and one of them above near the costa. The primaries are further ornamented by a series of small white translucent spots, bordered inwardly by blackish. These spots are arranged in a straight transverse series,

two in cell 1, one, transversely elongated, in cell 2, and a smaller
one in cell 3. The secondaries are heavily marked with black on
the outer angle, and there is a curved series of three or four small
black spots in the subcostal interspaces. Just after the large black
spot on the outer angle, the outer margin is lightly touched with
whitish. A fine dark marginal line defines the origin of the cilia,
which are fuscous upon the upperside. On the underside the
primaries are blackish, shading slightly into bluish grey at the
base. The translucent spots appear as on the upperside; the
two spots in cell 1 being defined outwardly by two parallel whitish
rays. The secondaries are white, laved with bluish grey at the
base. The outer angle is black. The black spots on the subcostal
interspaces are as on the upper surface, but more clearly defined
upon the white ground. In addition there are two small discal
dots in cell 1, and a small black dot on the outer margin near the
extremity of vein 1. The cilia on the underside are white toward
the anal angle.

Expanse 35 mm.

Hab. Gaboon (*Mocquerys*). Type in collection Staudinger.

18. S. ELIMINATA, sp. nov. (Plate V. fig. 9.)

♂. The colour of the upperside of the thorax and abdomen is
dark fuscous, of the underside yellowish ochraceous. The antennæ
are black, the legs grey, edged with blackish anteriorly. The primaries on the upperside are fuscous. There are three small
confluent subapical spots, a similar small spot on the upper edge
of the cell near its end, and two other like spots in cells 2 and 3,
of which the former is the larger. Both the subapical series and
the discal spots are followed inwardly by dark cloudings. The
interspaces just before the margin are marked by obscure darker
oblong spots. There is a fine dark marginal line. The cilia are
fuscous. The secondaries are traversed by a series of obscure dark
fuscous transverse median, limbal, and submarginal spots. The
spot of the median series located at the end of the cell is annuli-
form. The marginal line and cilia are as on the primaries. Both
the primaries and secondaries on the underside are clear yellowish
ochraceous, with the cilia pale fuscous. The inner margin of the
primaries is testaceous. The translucent spots of the upper surface
reappear upon the lower side and are narrowly margined with
fuscous. Fuscous submarginal and limbal bands traverse the
primaries, leaving sagittate spots of the prevailing ground-colour
between them on the intra-neural spaces. The secondaries show
the transverse series of spots of the upper surface, but more
distinctly defined and generally rounded than on the upperside.

♀ like the male.

Expanse 28–30 mm.

Hab. Abyssinia (*Staudinger*); Somaliland (*in my collection*).

This species is labelled in the Staudinger collection by Mons.
Mabille as "*nerva*, Fabr." Mons. Mabille has on several occasions
in his correspondence with me insisted upon employing the Fabri-

cian name for this insect. Perhaps he is following in this the
example of Ploetz, who referred some insect obtained from Kordofan
to the Fabrician species. But, whatever may have been the insect
before Ploetz at the time he was writing, it is certain that it was
not the insect described by Fabricius. In Jones's ' Icones ' (unpublished) we have the best clue to many of the Fabrician species, and
the figure of *H. nerva* there given (vide pl. 72, fig. 3) represents
undoubtedly a species of *Hesperia* (*Pyrgus*, Hübn. et auct.). The
published references to *Hesperia nerva*, Fabr., are the following:—

Hesperia nerva, Fabr. Ent. Syst. iii. p. 340, no. 293 (1793);
Latreille, Enc. Méth. ix. p. 789, no. 162 (1823).

Pyrgus nerva, Butl. Fabr. Diurn. Lep. p. 282 (1869).

Ephyriades nerva, Ploetz, JB. Nass. Ver. xxxvii. p. 6 (1884).

The habitat of *H. nerva* is given by Fabricius as " *in Indiis*," to
which little significance need be attached, as we know that this
phrase with the old writers often meant no more than that the
insect came from a foreign country.

19. S. AURIMARGO (Mab. MS.), sp. nov.   (Plate IV. fig. 8.)

Tabraca aurimargo, Mab. *in literis*.

♂. The antennœ and the upperside of the thorax and abdomen
are black, as is also the underside of the thorax and abdomen,
except at the anal extremity, where it is marked with orange-yellow;
the ground-colour of the primaries and secondaries is dark brown,
almost black. The primaries are ornamented by three minute
translucent subapical spots in the usual position. The outer
margin of the secondaries near the anal angle and the cilia for
the inner half of the wing are orange. On the underside, the
primaries are coloured and marked as upon the upperside. The
secondaries have the orange colour which appears upon the upperside near the anal angle much more broadly diffused, covering the
outer half of the wing as far as the subcostal nervules. The costal
margin and the base are broadly blackish brown, and the yellow
space is interrupted by an irregular row of discal spots, of which
the one opposite the end of the cell is the largest and confluent
with the dark costal area.

Expanse 28–30 mm.

Hab. Gaboon (*Mocquerys*); Sierra Leone (*Preuss*). Types in
coll. Staudinger.

This beautiful species has been named *Tabraca aurimargo* by
Mons. Mabille. In neuration and most other respects it agrees
with *Sarangesa* absolutely, and I cannot bring myself to recognize
in it the type of a new genus.

20. S. MACULATA, Mab.

Sape maculata, Mab. C. R. Soc. Ent. Belg. 1891, p. lxviii.

Hab. Mozambique (*Mabille*).

I have no clue to the determination of this species other than
the description of the author.

21. S. OPHTHALMICA, Mab.

Sape ophthalmica, Mab. C. R. Soc. Ent. Belg. 1891, p. lxviii.

Hab. Delagoa Bay (*Mabille*).

No specimen or figure of this species being available, I must content myself with a provisional reference to this location in the genus, to which the author has assigned it.

22. S. (?) PLISTONICUS, Ploetz.

Antigonus plistonicus, Ploetz, S. E. Z. vol. xl. p. 302 (1879).

Hab. Aburi (*Ploetz*).

I cannot make out this species from the description and the material before me. The description does not exactly apply to anything I have seen in nature, though it may be that it designates some already well-known species.

23. S. (?) PHILOTOMUS, Ploetz.

Antigonus philotomus, Ploetz, S. E. Z. vol. xl. p. 301 (1879); Karsch, Berl. Ent. Zeit. vol. xxxviii. p. 262 (1893).

Hab. Aburi (*Ploetz*); Togoland (*Karsch*).

I do not know this species, at least under this name.

24. S. (?) LÆLIUS, Mab.

Pterygospidea lælius, Ploetz MS., Mabille, Bull. Soc. Zool. France, 1877, p. 240

Ephyriades lælius, Ploetz, JB. Nass. Ver. xxxvii. p. 6.

Hab. Gaboon.

This is another species about which I am left in total uncertainty. Ploetz merely cites the name, and from the description of Mons. Mabille I cannot draw positive conclusions. Mons. Mabille has designated for me under this name two wholly different species, one being the species which he has labelled in the collection of Dr. Staudinger as *bouvieri*, and the other being a slight variety of *S. thecla*, Ploetz, which he named from a photographic representation sent to him, in which only the upperside appeared. I leave this puzzle somewhat reluctantly to others to solve.

25. S. KOBELA, Trim.

Nisoniades kobela, Trim. Trans. Ent. Soc. Lond. (3) vol. ii. p. 180 (1864); Rhop. Afr. Austr. ii. p. 312, pl. vi. fig. 4, *o* (1866).

Pterygospidea kobela, Trim. S. Afr. Butt. vol. iii. p. 353 (1889).

Sarangesa kobela, Watson, P. Z. S. 1893, p. 48.

Hab. Extra-tropical South Africa (*Trimen*).

This species reveals a striking superficial resemblance to the species of the genus *Thanaos*, and represents a section of the genus in which it stands thus far unique.

CELÆNORRHINUS, Hübn.

26. C. GALENUS, Fabr.

Hesperia galenus, Fabr. Ent. Syst. iii. 1, p. 350, no. 332 (1793);
Latr. Enc. Méth. ix. p. 773, no. 124 (1823).
Hesperia galena, Don. Ins. Ind. pl. 1. fig. 3, ♀ (1800).
Celænorrhinus galenus, Wats. P. Z. S. 1893, p. 49.
Plesioneura galenus, Staudgr. Exot. Schmett. pl. 100.
Pardaleodes fulgens, Mab. Bull. Soc. Zool. France, 1877,
p. 236, ♂.
Pterygospidea galenus, Trim. P. Z. S. 1894, p. 80.

Donovan in his plate figures the female of this species, which
may always be recognized by the elongate marginal spot on the
secondaries beyond the end of the cell. This spot has the form of
a parallelogram, and does not fuse with the adjacent spots so fully
as is the case in the male, where its sharp outlines are lost in the
spots on either side of it. Dr. Staudinger gives a good figure of
the male in his ' Exotische Schmetterlinge.' Mons. Mabille kindly
determined for me a number of species upon the occasion of a
recent visit to Paris, among them *Pardaleodes fulgens*, Mab. The
specimens so determined are undoubtedly *C. galenus*, Fabr., ♂. I
have a series of nearly 100 specimens of both sexes, some of them
taken *in coitu*, and am satisfied of the correctness of the synonymy
given as above.

This is one of the commonest of West-African butterflies and is
found from Senegambia to Upper Angola, and Manica (*Trimen*).

27. C. RUTILANS, Mab.

Pardaleodes rutilans, Mab. Bull. Soc. Zool. France, 1877, p. 235,
♀ ; Bull. Soc. Ent. France (Feb. 1877), ♀ ; Novit. Lepidopt.
p. 96, pl. xiii. fig. 7, ♂ (1893).
Pterygospidea tergemira, Hew. Ann. & Mag. Nat. Hist. (4)
vol. xx. p. 323 (Oct. 1877).
Tagiades woermanni, Ploetz, S. E. Z. vol. xl. p. 362, ♀ (1879).

Having seen the types of *P. rutilans*, Mab., and of *P. tergemira*,
Hew., and a carefully executed copy of the drawing of *T. woermanni*, ♀, made by Ploetz, I have not a shadow of doubt as to the
correctness of the above synonymy.

Hab. Fernando Po (*Hewitson*); Victoria, W. Africa (*Ploetz*);
Congo-Landana (*Mabille*); Gaboon, Cameroons (*Good*).

28. C. ILLUSTRIS, Mab.

Pardaleodes illustris, Mab. C. R. Soc. Ent. Belg. 1891, p. lxxiii.
Celænorrhinus illustris, Holl. Ent. News, March 1894, pl. iii.
fig. 6.

Hab. Cameroons and Upper Valley of the Ogové.

29. C. MEDETRINA, Hew. (Plate III. fig. 2.)

Pterygospidea medetrina, Hew. Ann. & Mag. Nat. Hist. (4)
vol. xx. p. 322 (1877).

Pardaleodes interniplaga, Mab. C. R. Soc. Ent. Belg. 1891,
p. lxxiii.
Celænorrhinus interniplaga, Holland, Ent. News, March 1894,
pl. iii. fig. 2.
Hab. Fernando Po (*Hewitson*); Cameroons (*Mabille*); Buló
Country (*Good*).
I am unable to discover any valid specific differences between
C. meditrina, Hew., and *C. interniplaga*, Mab. I have a good
series of specimens in my collection, some of which agree positively
with either form, differing only in size and the greater or less
distinctness of the marginal spots.

30. C. MACULATUS, Hampson.   (Plate III. fig. 4.)

Coladenia maculata, Hpsn. Ann. & Mag. Nat. Hist. (6) vol. vii.
p. 183.
Hab. Sabaki River, E. Africa (*Hampson*).
This species is a very near ally of *C. meditrina*, Hew. Two
specimens, a male and a female, contained in the collection of
Dr. Staudinger, were taken by Mocquerys at Gaboon. The
female differs from the male in having the maculations of the
secondaries greatly reduced in size. While these specimens do
not agree absolutely with the type of *maculata*, Hpsn., they are by
far too close to warrant a separation.

31. C. BISERIATUS, Butl.   (Plate III. fig. 3.)

Plesioneura biseriata, Butl. P. Z. S. 1888, p. 97.
Plesioneura hoehneli, Rogenhofer, Ann. Hofmus. Wien, vol. vi.
p. 463, pl. xv. fig. 10 (1891).
Hab. Kilimanjaro (*Butler*); Tropical Africa (*Rogenhofer*).
I think the above synonymy will be found to be quite correct.

32. C. ATRATUS, Mab.

Pardaleodes atratus, Mab. C. R. Soc. Ent. Belg. 1891, p. lxxiv.
Celænorrhinus collucens, Holl. Ent. News, March 1894, p. 90,
pl. iii. figs. 3, 4.
Hab. Cameroons (*Mabille*; *Good*).
The type of *P. atratus* being before me as I write, I am convinced that I made an error in my identification of it upon the
occasion of my visit to Mons. Mabille. The insect I labelled
atratus, if there has been no confusion since made in the labelling
of the specimens in the collection of Dr. Staudinger, is the
following species, and the true *atratus* is the species I figured and
named *collucens*. Dr. Staudinger warns me that Mons. Mabille
has in a few cases apparently confused his types: this is one of
those cases in which I am almost positive that such a confusion
has arisen; but we must accept the type as determining controversy,
and as the insect labelled autographically as *Pardaleodes atratus*
by Mabille in the Staudinger Collection is unmistakably my

collucens, and not the next species in this series, we must regard the identification as positively settled in this way.

33. C. BOADICEA, Hew.   (Plate III. fig. 1.)

Pterygospidea boadicea, Hew. Ann. & Mag. Nat. Hist. (4) vol. xx. p. 323 (1877).
Celænorrhinus atratus, Holl. Ent. News, March 1894, pl. iii. fig. 5.
Parduleodes lucens, Mab., MS.
Hab. Gaboon, Cameroons.
Mons. Mabille, in the 'Comptes Rendus de la Société Entomologique de Belgique,' 1891, p. lxxiv, in his description of *Parduleodes* (*Celænorrhinus*) *atratus*, alludes to a species of the genus named *lucens* by him from a figure of his type, which he has never published, so far as I am aware; I have been enabled to identify it with *boadicea*, Hew., which is undoubtedly the same insect figured by me in the 'Entomological News' for March 1894, as *C. atratus*, Mab. *C. boadicea*, Hew., may be distinguished from all other species by the greater breadth of the median yellow band on the primaries, and the larger expanse of the marginal spot near the outer angle of the secondaries on the upperside. This species is closely related to *C. atratus*, but quite distinct.

34. C. CHRYSOGLOSSA, Mab.   (Plate III. fig. 5.)

Ancistrocampta chrysoglossa, Mab. C. R. Soc. Ent. Belg. vol. xxxv. p. cvii (1891).

Hab. Cameroons (*Mabille*; *Good*).
The type of the species is a female. The figure in the Plate is taken from a male specimen in my collection. The insect undoubtedly is a *Celænorrhinus*, but differs from the other African species in being more plainly marked upon the primaries.

35. C. PROXIMUS, Mab.

Plesioneura proxima, Mab. Bull. Soc. Zool. France, 1877, p. 231; Ann. Soc. Ent. France, (6) vol. x. p. 31, pl. iii. fig. 1.
Tagiades elmina, Ploetz, S. E. Z. vol. xl. p. 362 (1879).
Hab. Gaboon, Cameroons, Sierra Leone, Togoland.

36. C. MACROSTICTUS, Holl.

C. macrostictus, Holl. Ent. News, Jan. 1894, p. 27, pl. i. fig. 2.
Hab. Valley of the Ogové.

37. C. HUMBLOTI, Mab.

Plesioneura humbloti, Mab. Bull. Soc. Ent. Belg. 1884, p. clxxxvii; Grandidier's Madagascar, vol. xiii. p. 349, pl. 54. figs. 8, 8 *a*.
Hab. Madagascar.

38. C. (?) HOMEYERI, Ploetz.

Tagiades homeyeri, Ploetz, S. E. Z. vol. xli. p. 307 (1880).

Hab. Pundo Ndongo.

I do not know this species, but as it is said by the author to bo very near *C. galenus*, Fabr., I locate it here provisionally.

39. C. MOKEEZI, Wallgr.

Pterygospidea mokeezi, Wallgr. K. Sv. Vet.-Acad. Handl. 1857; Lep. Rhop. Caffr. p. 54.

Hesperia amaponda, Trim. Trans. Ent. Soc. Lond. (3) vol. i. p. 405.

Nisoniades mokeezi, Trim. Rhop. Afr. Aust. vol. ii. p. 316, pl. vi. fig. 5.

Pterygospidea mokeezi, Trim. Butt. S. Afr. vol. iii. p. 358.

Celænorrhinus mokeezi, Watson, P. Z. S. 1893, p. 50.

Hab. Extra-tropical S. Africa.

40. C. (?) LUEHDERI, Ploetz.

Plastingia luehderi, Ploetz, S. E. Z. vol. xl. p. 357 (1879), vol. xlv. p. 147 (1884).

Hab. Aburi (*Ploetz*).

The figure of this species drawn by Ploetz appears to be a crude representation of a species of *Celænorrhinus*, but the statement of Ploetz, that there is a sexual mark or brand upon the primaries, does not agree with this view. I am at a loss, without having the insect before me, to say where it should be located. Mons. Mabille's note upon the drawing of Ploetz, contained in one of his manuscript comments upon the Ploetzian figures, strikes me as very appropriate, "*mihi non verisimile videtur.*"

TRICHOSEMEIA [1], Holl.

41. T. SUBOLIVESCENS, Holl. (Plate V. fig. 15.)

T. subolivescens, Holl. Ann. & Mag. Nat. Hist. Oct. 1892, p. 294; Wats. P. Z. S. 1893, p. 53.

Hab. Matabeleland.

42. T. TETRASTIGMA, Mab.

Ceratrichia tetrastigma, Mab. C. R. Soc. Ent. Belg. 1891, p. lxv; Novit. Lepidopt. p. 119, pl. xvi. fig. 8.

Hab. Interior of Cameroons (*Staudinger*).

Mons. Mabille refers this species with some doubt to the genus *Ceratrichia*. With his type before me, I am able to assert that the species is positively congeneric with the type of the genus *Trichosemeia*. It may even prove to be true that the two species are the same, in which case Mons. Mabille's name will have priority. There is, however, considerable difference in the colour

---

[1] By a typographical error, printed originally as "*Tricosemeia.*"

and markings of the underside of the secondaries, and it would
not be at all safe to merge the two forms under the same name
until we have more material.

43. T. QUATERNA, Mab.

Ceratrichia quaterna, Mab. C. R. Soc. Ent. France, 1889,
p. clvi; Novit. Lepidopt. p. 20, pl. iii. fig. 3 (1891).

Hab. Sierra Leone (*Mabille*).

This beautiful species, the type of which is before me as I write,
is correctly referred to the genus *Trichosemeia.*

44. T. (?) BRIGIDA, Ploetz.

Antigonus brigida, Ploetz, S. E. Z. vol. xl. p. 361 (1879).

Hab. Cameroons (*Good*); Roorke's Drift, S. Africa (*in my
collection*).

What I take to be the species named *brigida* by Ploetz is a
species which is more properly located in this genus than any
other at present constituted, though the secondaries lack the
characteristic hairy brand near the costa on the upperside, which
led me to give the name which I have applied to this genus.
This remark holds good also of the two following species.

45. T. (?) HEREUS, Druce. (Plate IV. fig. 21.)

Tagiades hereus, Druce, P. Z. S. 1875, p. 417.

Hab. Angola (*Monteiro*).

This species seems to be closely allied to, if not identical with,
S. brigida, Ploetz. In case of identity the name given by
Mr. Druce has priority.

46. T. (?) SUBALBIDA, Holl.

Sarangesa subalbida, Holl. Ent. News, Jan. 1894, p. 26, pl. i.
fig. 7.

Hab. Valley of the Ogové (*Good*).

In the form of the wings and the neuration, together with the
form of the antennæ, this species comes nearer those which are
strictly classified in the genus *Trichosemeia* than to those included
in *Sarangesa.* The hairy brand on the upperside of the secondaries
is lacking; but in spite of this I prefer to place the species here,
rather than to leave it where I originally located it.

TAGIADES, Hübn.

47. T. FLESUS, Fabr.

Papilio flesus, Fabr. Spec. Ins. ii. p. 135, no. 621 (1871) ; Mant.
Ins. p. 88, no. 797 (1787) ; Ent. Syst. iii. p. 338, no. 286 (1793).

Nisoniades flesus, Butl. Cat. Fabr. Diurn. Lep. p. 286.

Papilio ophion, Dru. Ill. Exot. Ent. vol. iii. pl. xvii. figs. 1, 2
(1782); Stoll, Suppl. Cram. Pap. Exot. p. 127, pl. xxvi. figs. 4, 4 c
(1791).

Nisoniades ophion, Trim. Rhop. Afr. Aust. vol. ii. p. 313 (1866).

Pterygospidea flesus, Trim. Butt. S. Afr. vol. iii. p. 363 (1889).
Tagiades flesus, Wats. P. Z. S. 1893, p. 54.

Hab. Africa, south of the Sahara.

48. T. INSULARIS, Mab.

T. insularis, Mab. Ann. Soc. Eut. France, 1876, p. 272;
Grandidier's Madagascar, vol. xiii. p. 352 pl. 54. figs. 6, 7, 7 a.
Thymele ophion, Boisd. Faune Entomol. Madgr. p. 62, pl. ix.
fig. 4 (1833).

Hab. Madagascar.

This is the insular form of *T. flesus*, Fabr., which is found in
Madagascar, and can scarcely be separated from the Fabrician
species.

49. T. LACTEUS, Mab.

Tagiades lacteus, Mab. Bull. Soc. Ent. France, (5) vol. vii.
p. xxxix.
Tagiades dannatti, Ehrmann, Ent. News, vol. iv. p. 309; Holl.
Ent. News, March 1894, pl. iii. fig. 1.

Hab. Congo, Liberia.

My surmise that *T. lacteus* and *T. dannatti* are identical, which
I expressed in my paper of March 1894, has been confirmed by
Mons. Mabille, who has compared my figure with the type.

50. T. SAMBORANA, H. G. Smith.

Tagiades samborana, H. Grose Smith, Ann. & Mag. Nat. Hist.
(6) vol. vii. p. 127.

Hab. Madagascar.

I do not know this species.

51. T. SMITHII, Mab.

Tagiades smithii, Mab. Grandidier's Madagascar, vol. xiii.
p. 354, pl. 56 A. figs. 3, 3 a.

Hab. Madagascar.

The plate on which this species is to be figured has not yet been
published. I do not know the species in nature, nor by any
pictorial representation.

EAGRIS, Guen.

52. E. SABADIUS, Gray.

Hesperia sabadius, Gray, Griff. An. Kingd. vol. xv. pl. 99. fig. 2
(1832).
Thymele sabadius, Boisd. Faun. Eutom. Madgr. p. 63, pl. ix.
fig. 2 (1833).
Eagris sabadius, Guen. Maill. Réun. vol. ii. Lép. p. 18 (1863);
Mab. Grandid. Madagr. vol. xiii. p. 350, pl. 54. figs. 4, 4 a, 5.
Hesperia andrachne, Boisd. Faun. Eut. Madgr. p. 67 (1833);
Guérin, Iconogr. Règne Anim., Ins. pl. lxxxii. fig. 2 (1844).
Antigonus andrachne, Saalm. Lep. Madgr. p. 112, pl. i. fig. 14.

Plesioneura hyalinata, Saalm. Ber. Senck. Ges. 1877-78, p. 87.
Plesioneura andrachne (Boisd.), Saalm. Ber. Senck. Ges. 1878-79,
p. 123.
Hab. Madagascar.

53. E. NOTTOANA, Wallgr.

Pterygospidea nottoana, Wallgr. K. Sv. Vet.-Akad. Handl. 1857 ;
Lep. Rhop. Caffr. p. 54.
Nisoniades sabadius, Trim. Rhop. Afr. Austr. vol. ii. p. 315.
Pterygospidea nottoana, Trim. S. Afr. Butt. vol. iii. p. 360.
Eagris melancholica, Mab. Bull. Soc. Ent. France, (6) vol. ix.
p. clv.
Eagris nottoana, Wats. P. Z. S. 1893, p. 54.
Hab. South Africa.
The comparison of the type of *E. melancholica*, Mab., shows it to
be identical with *E. nottoana*, as determined by Mr. Trimen.

54. E. DECASTIGMA, Mab.

Eagris decastigma, Mab. C. R. Soc. Ent. Belg. 1891, p. lxii ;
Novit. Lepidopt. p. 118, pl. xvi. fig. 7 ; Holl. Ent. News, Jan.
1894, pl. i. fig. 9.
Hab. Sierra Leone, Gaboon.

55. E. FUSCOSA, Holl.   (Plate V. fig. 4.)

Eagris fuscosa, Holl. Ent. News, Jan. 1894, p. 27, pl. i. fig. 6.
Hab. Valley of the Ogové (*Good*); Gaboon (*Mocquerys*).
This is a somewhat close ally of *E. phyllophila*, Trim., but may
be readily distinguished from that species by the form of the large
spots on the disk of the primaries.

56. E. PHYLLOPHILA, Trim.

Nisoniades phyllophila, Trim. Trans. Ent. Soc. Lond. 1883,
p. 362.
Pterygospidea phyllophila, Trim. S. Afr. Butt. vol. iii. p. 362,
pl. xii. fig. 8.
Hab. Natal, Delagoa Bay (*Trimen*).

57. E. JAMESONI, Sharpe.

Antigonus jamesoni, Sharpe, Ann. & Mag. Nat. Hist. (6) vol. vi.
p. 348 (1890).
Pterygospidea jamesoni, Trim. P. Z. S. 1891, p. 106, pl. ix. fig. 25.
Caprona jamesoni, Butl. P. Z. S. 1893, p. 669.
Hab. S.W. Africa, Mashonaland.

58. E. DENUBA, Ploetz.   (Plate V. fig. 8.)

Antigonus denuba, Ploetz, S. E. Z. vol. xl. p. 361 (1869).
Eagris decolor, Mab. Bull. Soc. Ent. France, (6) vol. ix. p. clv
(1889); Karsch, Berl. Ent. Zeit. vol. xxxviii. p. 262 (1893).

Hab. Aburi (*Ploetz*); Freetown (*Mabille*); Cameroons (*Good*); Togoland (*Karsch*).

Having before me a drawing of the type of Ploetz, executed by Prillwitz, which is reproduced in the Plate, and the type of Mabille, loaned me by Dr. Staudinger, I am positively satisfied as to the identity of the two.

59. E. LUCETIA, Hew.

Leucochitonea lucetia, Hew. Ill. Exot. Butt. vol. v. Hesp., *Leucochitonea*, pl. ii. fig. 21.

Hab. Angola (*Hewitson*).

PROCAMPTA, Holl.

60. P. RARA, Holl.

Procampta rara, Holl. Ann. & Mag. Nat. Hist. Oct. 1892, p. 293; Watson, P. Z. S. 1893, p. 59; Holl. Ent. News, Mar. 1894, pl. iii. fig. 7.

Hab. Valley of the Ogové.

CAPRONA, Wallgr.

61. C. PILLAANA, Wallgr.

Caprona pillaana, Wallgr. K. Sv. Vet.-Akad. Handl. 1857; Lep. Rhop. Caffr. p. 51; Trim. Rhop. Afr. Austr. vol. ii. p. 308 (1866); S. Afr. Butt. vol. iii. p. 348, pl. xii. figs. 6, 6 *a* (1889).

Stethotrix heterogyna, Mab. Bull. Soc. Ent. France, (6) vol. ix. p. clxxxiv (1889).

Caprona adelica, Karsch, Ent. Nachr. vol. xviii. p. 242 (1892); Berl. Ent. Zeit. xxxviii. p. 243, pl. vi. fig. 2 (1893).

Hab. South Africa, Natal, Loko, Togoland.

Mons. Mabille writes me that the species of Karsch is absolutely identical with his *S. heterogyna*, in which opinion, with the type before me as I write, I am able to positively concur. But the male of *S. heterogyna* is most certainly identical with *C. pillaana*, Wallgr. I am not alone in this opinion. Dr. Staudinger writes me that Prof. Aurivillius has most unqualifiedly given in his adhesion to this view on examination of specimens submitted to him. The female, the type of which is before me, might have served the artist for the drawing of *C. adelica* given by Dr. Karsch, and differs from the rather crude figure of the female of *C. pillaana*, Wallgr., given by Trimen in being paler, and having a sharply defined black spot on the underside of the secondaries near the inner margin. With only the female sex before me I might have hesitated a little to make the above synonymy, but the identity of the male with *C. pillaana* being so positively certain, I do not doubt the correctness of what I have given above.

62. C. CANOPUS, Trim.

Caprona canopus, Trim. Trans. Ent. Soc. Lond. (3) vol. ii. p. 180

2*

(1864); Rhop. Afr. Austr. vol. ii. p. 309, pl. vi. fig. 2 (1866);
Staud. Exot. Schmett. pl. 100 ; Trim. S. Afr. Butt. p. 349 (1889).
Hab. Extra-tropical South Africa.

ABANTIS, Hopff.

(*Leucochitonea*, Wallgr. ; *Sapœa*, Ploetz.)

I cannot bring myself to differ from Trimen, and to accept the
conclusion of Watson, that *L. levubu*, Wallgr., should constitute the
type and sole representative of a genus. The difference between
this species and the others given below are certainly rather of
specific than of generic grade. I therefore sink Wallengren's
genus *Leucochitonea* as a synonym of *Abantis*, Hopff., as has
already been done by Trimen.

63. A. TETTENSIS, Hopff.

Abantis tettensis, Hopff. Mouatsb. k. Akad. Wiss. Berl. 1855,
p. 643 ; Peters' Reise Mossamb., Ins. p. 415, pl. xxvi. figs. 10, 17
(1862); Trim. S. Afr. Butt. vol. iii. p. 337 (1889); Wats. P. Z. S.
1893, p. 63.
Hab. South Tropical and Temperate Africa.

64. A. PARADISEA, Butl.

Leucochitonea paradisea, Butl. Trans. Ent. Soc. Lond. 1870,
p. 499 ; Lep. Exot. p. 167, pl. lix. fig. 8 (1874).
Hesperia (Oxynetra) namaquana, Westw. Thes. Ent. Oxon.
p. 183, pl. xxxiv. fig. 10 (1874).
Leucochitonea paradisea, Staudgr. Exot. Schmett. i. pl. 100.
Abantis paradisea, Trim. S. Afr. Butt. p. 342 (1889) ; Wats.
P. Z. S. 1893, p. 63.
Hab. Southern Africa.

65. A. ZAMBESIACA, Westw.

Hesperia zambesiaca, Westw. Thes. Ent. Oxon. p. 183, pl. xxxiv.
fig. 9 (1874).
Abantis zambesina, Trim. S. Afr. Butt. vol. iii. p. 344 (1889);
P. Z. S. 1891, p. 105.
Sapœa trimeni, Butl. P. Z. S. 1895, p. 264, pl. xv. fig. 5.
Hab. Southern Tropical Africa.

With the figures of their species, given by Westwood and
Butler, before me and a long series of specimens labelled by
Mr. Trimen to compare with them, I am wholly at a loss to see
what valid reason exists for separating the insect recognized by
Dr. Butler as *Sapœa trimeni* from the insect described by Westwood. It is true that the normal colour of the sides of the
abdominal segments of the insect is " snow-white," as stated by
Dr. Butler, and brought out in his excellent figure, but the fact
that Westwood says that these segments in the type were
' luteous " does not in my judgment furnish sufficient reason to

say that we are dealing here with two distinct species. "Luteous" is muddy yellow, and nothing is commoner among the Hesperiidæ than the change of the white markings of the abdomen into yellowish by greasing and other accidents. I am reluctant to differ from my learned friend Dr. Butler on any point, but after studying the specimens before me with the figures and descriptions given by himself and Westwood, I am still of the opinion that Mr. Trimen's original identification was correct, and that the separation of the form known to Trimen from that described by Westwood is an unnecessary refinement.

66. A. BISMARKI, Karsch.

Abantis bismarki, Karsch, Ent. Nachr. xviii. p. 228 (1892); Berl. Ent. Zeit. vol. xxxviii. p. 242, pl. vi. fig. 1 (1893).
Hab. Togoland.

67. A. BICOLOR, Trim.

Leucochitonea bicolor, Trim. Trans. Ent. Soc. Lond. (3) vol. ii. p. 180 (1864); Rhop. Afr. Austr. vol. ii. p. 307, pl. 6. fig. 1 (1866).
Sapæa bicolor, Ploetz, S. E. Z. vol. xl. pp. 177, 179 (1879).
Abantis bicolor, Trim. S. Afr. Butt. vol. iii. p. 340 (1889); Wats. P. Z. S. 1893, p. 63.
Hab. S. Africa.

68. A. VENOSA, Trim.

Abantis venosa, Trim. S. Afr. Butt. vol. iii. p. 339 (1889); P. Z. S. 1891, p. 105, pl. ix. fig. 24.
Leucochitonea umvulensis, Sharpe, Ann. & Mag. Nat. Hist. (6) vol. vi. p. 348 (1890).
Hab. South Tropical Africa and Transvaal.

69. A. ELEGANTULA, Mab.

Abantis elegantula, Mab. Ann. Soc. Ent. France, 1890, p. 32; Novit. Lepidopt. p. 23, pl. iii. fig. 6 (1891).
Hab. Sierra Leone.

70. A. EFULENSIS, sp. nov. (Plate V. fig. 12.)

♂. Allied to *A. elegantula*, Mab., from which it differs by the entire absence of the discal spots on the primaries. The secondaries are white, with the basal third, the outer angle, and the inner margin clouded with dark brown, shading on the costa into orange-red. The white outer area is intersected by the veins, which are black.

On the underside, the primaries are much paler than on the upperside and are slightly tinged near the base and on the costa with ochreous. The secondaries are pure white, except on the costal margin and the outer angle, where they are laved with pale brown shading into ochraceous. The veins on the underside are not black as on the upperside, except those which are located near the costa. The body is marked much as in *A. elegantula*, but is without the red spots at the end of the patagia and the red hairs which are found on the metathorax. Expanse 40 mm.

Hab. Efulen, Cameroons.

71. A. LEUCOGASTER, Mab.

Abantis leucogaster, Mab. Ann. Soc. Ent. France, 1890, p. 32; Novit. Lepidopt. p. 22, pl. iii. fig. 5 (1891).

Hab. Sierra Leone.

72. A. LEVUBU, Wallgr.

Leucochitonea levubu, Wallgr. K. Sv. Vet.-Akad. Handl. 1857; Lep. Rhop. Caffr. p. 52; Trim. Rhop. Afr. Austr. vol. ii. p. 306. *Abantis levubu*, Trim. S. Afr. Butt. vol. iii. p. 345, pl. xii. fig. 5. *Hab.* Southern Africa.

HESPERIA, Fabr.

(*Pyrgus*, Hübn.; *Scelothrix*, Ramb.; *Syrichtus*, Boisd.)

73. H. SPIO, Linn.

Papilio spio, Linn. Syst. Nat. ed. xii. p. 796, no. 271 (1767); Fabr. Syst. Ent. p. 535, no. 400 (1775); Donovan, Ius. Ind. pl. i. fig. 5 (1800-3).
Hesperia spio, Fabr. Ent. Syst. iii. 1, p. 354, no. 348 (1783); Westw., Don. Ins. Ind. 2nd edit. p. 79, pl. 50. fig. 5 (1842); Aurivillius, K. Sv. Vet.-Akad. Handl. vol. xix. no. 5, p. 124, tab. i. figs. 3, 3 a, after Clerck (1882).
Papilio vindex, Cram. Pap. Exot. vol. iv. pl. cccliii. figs. G, H (1782); Watson, P. Z. S. 1893, p. 65.
Pyrgus vindex, Hübn. Verz. p. 109, no. 1178 (1810); Hopff. Peters' Reise Mossamb., Ius. p. 421 (1862); Trim. Rhop. Afr. Austr. vol. ii. p. 287 (1866); S. Afr. Butt. vol. iii. p. 280 (1889).
Hesperia vindex, Latr. Enc. Méth. vol. ix. p. 785 (1823); Westw., Doubl. & Hew. Gen. Diurn. Lep. pl. lxxix. fig. 6 (1852).
Syrichtus vindex, Wallgr. Rhop. Caffr. p. 53 (1857).

Hab. Southern Africa.

I had long been led to question whether this species had been found in the western tropical parts of Africa. I have never received it from Gaboon, Cameroons, Sierra Leone, or Liberia, though I have charged my collectors to make special search for the Hesperiidæ, and have received thousands of specimens from them. The species identified for me as *H. spio*, L. (*vindex*, Cram.), by several European authorities, is very different from the S.-African insect, of which I have numerous examples received from Mr. Trimen and others. It is *H. ploetzi*, Auriv. My doubt as to the existence of the species on the Tropical West Coast has been, however, put to rest by the discovery of a specimen from Monrovia in the collection of Dr. Staudinger.

74. H. DROMUS, Ploetz.

Pyrgus dromus, Ploetz, Mitth. nat. Ver. Neu-Vorpomm. u. Rüg.
1884, p. 6; Trim. S. Afr. Butt. vol. iii. p. 283 (1889).
Hesperia dromus, Watson, P. Z. S. 1893, p. 65.
Hab. South Africa (? North of the Congo).

This species is generally confounded in collections with the
preceding, but by attending to the differences so clearly pointed
out by Mr. Trimen they may easily be separated. Ploetz states
that his type was from the Congo, and Mr. Trimen, upon the
authority of G. Geynet, gives the " Gaboon River " as a habitat.
I am inclined to question the correctness of the reference of this
species to these localities. I may be in error, but am inclined to
think that it does not range further north than Angola on the
West Coast.

75. H. PLOETZI, Auriv.

Syrichtus spio, Ploetz, Mitth. nat. Ver. Neu-Vorpomm. u. Rüg.
1884, p. 21.
Pyrgus spio, Mab. Ann. Soc. Ent. France, (6) vol. x. p. 30, pl. iii.
fig. 9 (1890).
Hesperia ploetzi, Auriv. Ent. Tidsk. 1891, p. 227.
Pyrgus ploetzi, Karsch, Berl. Ent. Zeit. vol. xxxviii. p. 245
(1893).
Hab. Gaboon, Liberia, Sierra Leone, Togoland.

76. H. SATASPES, Trim.

Pyrgus sataspes, Trim. Trans. Ent. Soc. Lond. (3) vol. ii. p. 178
(1864); Rhop. Afr. Austr. vol. ii. p. 290, pl. v. fig. 7 (1866);
S. Afr. Butt. vol. iii. p. 289 (1889).
Hab. South Africa.

77. H. DIOMUS, Hopff.

Pyrgus diomus, Hopff. Monatsber. k. Akad. Wissensch. Berl.
1855, p. 643; Peters' Reise u. Mossamb., Ins. p. 420, pl. xxvii.
figs. 9, 10 (1862).
Hab. Tropical East Africa.

78. H. FEROX, Wallgr.

Syrichthus ferox, Wallgr. Wien. Ent. Monatschr. 1863, p. 137.
Pyrgus vindex, Cram.? var., Trim. Rhop. Afr. Austr. vol. ii.
pp. 287–288 (1866).
Hesperia (Syrichthus) diomus, Wallgr. Sv. Vet.-Akad. Förh
1872, p. 50.
Pyrgus diomus, Möschl. Verh. zool.-bot. Ges. Wien, 1883, p. 286.
Hesperia sandaster, Staudgr. Exot. Schmett. vol. ii. pl. 100
(1888).
Pyrgus diomus, Trim. S. Afr. Butt. vol. iii. p. 287 (1889).
Hesperia diomus, Wats. P. Z. S. 1893, p. 65.
Hab. Southern Africa.

" I have come to the conclusion that *Pyrgus diomus*, Hopff., is really distinct from *P. ferox*, Wallgr., although Wallengren himself in 1872 sank the latter in favour of the former. None of the South-African specimens that I have seen agrees with Hopffer's description and figures in the important point of the white bands on the underside of the hind wings, which markings are always much more oblique in the southern examples. The other day I received a pair from Zanzibar, which exactly agree with Hopffer's figures. So I think we may call the abundant southern form *P. ferox*. By the way, what Dr. Staudinger figures as my *P. sandaster* is apparently *P. ferox*." (R. Trimen, *in literis*, 1894.)

79. H. ASTERODIA, Trim.

Pyrgus asterodia, Trim. Trans. Ent. Soc. Lond. (3) vol. ii. p. 178 (1864); Rhop. Afr. Austr. vol. ii. p. 289, pl. v. fig. 6 (1866); S. Afr. Butt. vol. iii. p. 284 (1889).
Hesperia asterodia, Watson, P. Z. S. 1893, p. 65.
Syrichthus asterodia, Ploetz, Mitth. nat. Ver. Neu-Vorpomm. u. Rüg. 1884, p. 21.
Hab. South Africa.

80. H. TRANSVAALIÆ, Trim.

Pyrgus transvaaliæ, Trim. S. Afr. Butt. vol. iii. p. 286 (1889).
Hab. South Africa.
Allied, according to the author, to *H. spio*, Linn. (*vindex*, Cram.), and *dromus*, Ploetz.

81. H. AGYLLA, Trim.

Pyrgus agylla, Trim. S. Afr. Butt. vol. iii. p. 286 (1889).
Hab. South Africa.
This species is unknown to me except by the description of Mr. Trimen.

82. H. MAFA, Trim.

Pyrgus mafa, Trim. Trans. Ent. Soc. Lond. 1870, p. 386, pl. vi. fig. 12; S. Afr. Butt. vol. iii. p. 284.
Hab. South Africa.
Doubtfully distinct from *H. spio*, Linn.

83. H. SANDASTER, Trim.

Pyrgus sandaster, Trim. Trans. Ent. Soc. Lond. 1868, p. 92, pl. v. fig. 9; S. Afr. Butt. vol. iii. p. 291 (1889).
Hab. South Africa.

84. H. NANUS, Trim.

Pyrgus sataspes, var. A. Trim. Rhop. Afr. Austr. vol. ii. p. 290 (1866).
Pyrgus nanus, Trim. S. Afr. Butt. vol. iii. p. 290 (1889).
Hab. South Africa.

85. II. SECESSUS, Trim.

Pyrgus secessus, Trim. P. Z. S. 1891, p. 102, pl. ix. fig. 22.
Hab. South-western Africa.

86. H. COLOTES, Druce.   (Plate I. fig. 11.)

Pyrgus colotes, Druce, P. Z. S. 1875, p. 416.
Hab. Angola (*Monteiro*).

87. II. NORA, Ploetz.

Pyrgus nora, Ploetz, Mitth. nat. Ver. Neu-Vorpomm. u. Rüg.
1884, p. 7.
Hab. Loango (*Ploetz*).
This species is unknown to me, and may be identical with some
other species. The description is very unsatisfactory. In some
respects it applies to *H. secessus*, Trim.

88. II. ZAIRA, Ploetz.

Pyrgus zaira, Ploetz, Mitth. nat. Ver. Neu-Vorpomm. u. Rüg.
1884, p. 6.
Hab. Congo (*Ploetz*).
This species is only known to me by the brief and unsatisfactory
description of Ploetz.

89. II. ABSCONDITA, Ploetz.

Syrichthus abscondita, Ploetz, Mitth. nat. Ver. Neu-Vorpomm.
u. Rüg. 1884, p. 21.
Hab. Africa (*Ploetz*).
The description is too slight to base any conjecture upon it as
to what the author intended thereby.

90. II. PROTO, Esp.

Papilio proto, Esp. Eur. Schmett. i. 2, pl. 123. figs. 5, 6 (1806 ?).
For synonymy *cf.* Staudinger and Wocke, Kirby, Syn. Catalogue,
&c.
Hab. Morocco.

91. II. ALI, Oberth.

Syrichthus ali, Oberth. Étud. Entom. vi. 3, p. 61, pl. ii. fig. 3
(1881).
Hab. Algeria.

92. II. LEUZEÆ, Oberth.

Syrichthus leuzeæ, Oberth. Étud. Entom. vi. 3, p. 60, pl. iii.
fig. 10 (1881).
Hab. Algeria.

93. H. ALVEUS, Hübn., var. ONOPORDI, Ramb.

Syrichthus onopordi, Ramb. Faun. And. pl. viii. fig. 13 (1839).

For fuller synonymy *cf.* standard works on the Lepidoptera of the palæarctic faunal region.

Hab. North Africa.

94. H. (?) OILEUS, Linn.

Papilio oileus, Linn. Syst. Nat. i. 2, p. 795, no. 269 (1767).
Hesperia oileus, Kirby, Syn. Cat. p. 615 (1871).

Hab. Algeria (*Kirby*).

This is a doubtful species, and it does not appear that any one has been able to discover exactly what Linnæus intended to designate by his name and description. *Nominis umbra!!*

CARCHARODUS, Hübn.

(*Urbanus*, Hübn.; *Spilothyrus*, Dup.)

95. C. ALCEÆ, Esp.

Papilio alceæ, Eur. Schmett. i. 2, pl. li. fig. 3 (1780).

For further synonymy see standard works on the Lepidoptera of the palæarctic faunal region.

Hab. North Africa.

96. C. ELMA, Trim.

Pyrgus elma, Trim. Trans. Ent. Soc. Lond. (3) vol. i. p. 288 (1862); Rhop. Afr. Austr. vol. ii. p. 291, pl. v. fig. 8 (1866); S. Afr. Butt. vol. iii. p. 293.
Gomalia elma, Watson, P. Z. S. 1893, p. 67.
Pyrgus elma, Karsch, Berl. Ent. Zeit. vol. xxxviii. p. 245, pl. vi. fig. 12.

Hab. Southern Africa.

I place this insect in the genus *Carcharodus*, Hübn., rather than in the genus *Gomalia*, Moore, to which it has been assigned by Mr. Watson, because the differences of a structural character which separate it from its near allies, *C. alceæ* and *C. lavateræ*, are, in my opinion, too slight to warrant the subdivision. In fact, I call in question the propriety of retaining the name *Gomalia* as a generic designation, it being founded upon differences which appear to me to be rather specific than generic. I am quite persuaded that *Gomalia albofasciata*, Moore, the type of his genus, belongs to the older genus of Hübner, and I think *Gomalia* should be sunk as a synonym of *Carcharodus*.

The figure given by Karsch is by no means characteristic. The checkered character of the fringes is not made to appear, and were not the identification made by Karsch so positive, I should think we were dealing with some other species, belonging, perhaps, to a different genus.

97. C. (?) MIDEA, Walk.

Pelopidas midea, Walk. Entomologist, vol. v. p. 56 (1870).
Erynnis? midea, Kirby, Syn. Cat. p. 830 (1877).
Hab. Cairo.

I know nothing of this species. Mr. Kirby's reference to *Erynnis* leads me to place it here. Mr. Butler could not find the type in the British Museum. I fear that in this, as in so many other cases, we shall never be able to know exactly what Mr. Walker intended by his specific appellation.

Subfam. PAMPHILINÆ.

TRAPEZITES, Hübn.

The following species, all but one occurring in Madagascar, I allow to remain in the genus *Trapezites*, where they have been for the most part located by Dr. Butler and Mons. Mabille. Lieut. Watson states that the genus *Trapezites*, in the strict sense, is confined to the Australian region. Unfortunately I have not sufficient material at hand to justify the attempt by dissection and bleaching to determine whether these species are really separable from the genus in which they have hitherto been placed. It is much to be wished that some capable collector, who has an eye for the more obscure forms, might soon visit and thoroughly explore the field which is awaiting his labour in the great island east of Africa.

98. T. EMPYREUS, Mab.

Cyclopides empyreus, Mab. Pet. Nouv. Entom. vol. ii. p. 285 (1878).
Trapezites empyreus, Mab. Grand. Madgr. vol. xviii. p. 336, pl. liii. figs. 1, 1 a, 2 (1887).
Hab. Madagascar.

99. T. FASTUOSUS, Mab.

Cyclopides empyreus (pro parte), Mab. Pet. Nouv. Entom. vol. ii. p. 285 (1878).
Trapezites fastuosus, Mab. C. R. Soc. Ent. Belg. vol. xxviii. p. clxxxvi (1884); Graudid. Madgr. vol. xviii. p. 338, pl. liii. figs. 9, 9 a (1887).
Hab. Madagascar.

100. T. CARMIDES, Hew.

Cyclopides carmides, Hew. Descript. One Hundred New Hesperid. p. 41 (1868); Exot. Butt. vol. v. pl. *Cyclopides*, fig. 1 (1874).
Trapezites carmides, Mab. Grandid. Madgr. vol. xviii. p. 332, pl. liii. figs. 3, 3 a (1887).
Hab. Madagascar.

101. T. MALCHUS, Mab.

Cyclopides malchus, Mab. Bull. de la Soc. Philomat. p. 136 (1877).
Hesperia ypsilon, Saalm. Lep. Madgr. p. 110 (1884).
Trapezites malchus, Mab. Grandid. Madgr. vol. xviii. p. 332, pl. liii. figs. 5, 6 (1887).
Hab. Madagascar.

102. T. GILLIAS, Mab.

Pamphila gillias, Mab. Pet. Nouv. Entom. vol. ii. p. 285 (1878).
Trapezites kingdoni, Butl. Ann. & Mag. N. H. (5) vol. iv. p. 232 (1879).
Trapezites gillias, Mab. Grandid. Madgr. vol. xviii. p. 335, pl. liii. figs. 8, 8 a (1887).
Hab. Madagascar.

103. T. HOVA, Mab.

Cyclopides howa (err.), Mab. Bull. Soc. Ent. France, (5) vol. v. p. ccxv (1875).
Trapezites hova, Mab. Grandid. Madgr. vol. xviii. p. 335, pl. liii. figs. 7, 7 a (1887).
Hab. Madagascar.

104. T. CATOCALINUS, Mab.

Cyclopides catocalinus, Mab. Pet. Nouv. Entom. vol. ii. p. 285 (1878).
Trapezites catocalinus, Mab. Grandid. Madgr. vol. xviii. p. 339, pl. liii. figs. 4, 4 a (1887).
Hab. Madagascar. (Erroneously? labelled in Dr. Staudinger's collection as from the Gold Coast.)

105. T. PAROECHUS, Mab.

Trapezites paroechus, Mab. Grandid. Madgr. vol. xviii. p. 334, pl. lii. figs. 1, 2, 2 a (1887).
Hab. Madagascar.

106. T. (?) CHIRALA, Trim.

Pamphila chirala, Trim. P. Z. S. 1894, p. 76, pl. vi. fig. 18, ♀.

I place this species here provisionally, as, both from the figure and the description, it seems more nearly allied to the species in this group than to any others.

ACLEROS, Mab.

107. A. LEUCOPYGA, Mab.

Cyclopides leucopyga, Mab. Bull. Soc. Ent. France, 1877, p. 101.

Acleros leucopyga, Mab. Grandid. Madgr. vol. xiii. p. 347, pl. liv. figs. 3, 3 *a* (1887) ; Watson, P. Z. S. 1893, p. 76.

Hab. Madagascar.

This species may be distinguished from *A. ploetzi*, its near ally, by the broader extent of the white markings upon the outer margin of the secondaries, and the paler, more irregularly clouded underside of the secondaries. There are two specimens in the collection of Dr. Staudinger labelled as taken at Gaboon by Mocquerys, which are almost identical with examples from Madagascar. They were taken in September. (Are the locality-labels correct in these cases ?)

108. A. PLOETZI, Mab. (Plate II. fig. 7.)

Apaustus leucopygus, Ploetz, S. E. Z. vol. xl. p. 360 (1879).
Acleros ploetzi, Mab. Bull. Soc. Ent. France, (6) vol. ix. p. clxviii (1889).

Hab. Aburi, Victoria, W. Africa (*Ploetz*); Gaboon, Cameroons (*Good*).

Mons. Mabille has very properly suggested the name *ploetzi* for this species, in view of the fact that the specific name *leucopyga* had already, in 1877, been applied by him to a closely allied species from Madagascar.

109. A. MACKENII, Trim.

Pamphila ? *mackenii*, Trim. Trans. Ent. Soc. Lond. 1868, p. 95, pl. vi. fig. 8.
Ancyloxypha mackenii, Trim. S. Afr. Butt. vol. iii. p. 331 (1889).

Hab. Southern Africa.

This species is very closely allied to *A. ploetzi*, Mab. (*leucopygus*, Ploetz), but may be distinguished by its somewhat larger size, and by the fact that the underside of the primaries is much darker, and by the two subtriangular spots of white standing out boldly upon this dark ground near the inner margin.

110. A. PLACIDUS, Ploetz. (Plate II. fig. 10.)

Apaustus placidus, Ploetz, S. E. Z. vol. xl. p. 360 (1879), vol. xlv. p. 157 (1884).

Hab. Aburi (*Ploetz*).

The figure of the type given in the plate accompanying this article suggests that the original specimen is somewhat faded. I am greatly inclined to the view that it represents a somewhat rubbed specimen of the species since named *A. biguttulus* by Mons. Mabille, and which may also be identical with the species named *A. substrigata* by me. In a very long series of specimens, numbering nearly one hundred, I find specimens more or less worn, which agree well with the figure and description of *placidus*, and others which are undoubtedly very close to, if not identical with, *biguttulus*, and still others, bright and fresh, which are unmistakably separated from the others by the markings of the underside of the secondaries as represented in the photographic representation of *substrigata* given by me in the 'Entomological News' for January, 1894. Whether all of these belong to one and the same species remains to be proved, but the presumption seems to me to be in favour of this view. I do not, however, sink Mabille's species and my own as synonyms of *placidus* in the present paper, although inclined strongly to take this view.

111. A. DIGUTTULUS, Mab.

Acleros biguttulus, Mab. Bull. Soc. Ent. France, (6) vol. ix. p. clxvii (1889).

Hab. Freetown, W. Africa (*Mabille*).

From the brief description of the species given by Mons. Mabille, this species appears to me to be very near *A. placidus*, Ploetz (*q. v.*).

112. A. SUBSTRIGATA, Holl.

Acleros substrigata, Holl. Ent. News, Jan. 1894, p. 28, pl. i. figs. 10, 11.

Hab. Valley of the Ogové.

This is possibly a form of *placidus*, Ploetz, as I have intimated above.

113. A. OLAUS, Ploetz.

Apaustus olaus, Ploetz, S. E. Z. vol. xlv. p. 156 (1884); Karsch, Berl. Ent. Zeit. 1893, p. 260.

Hab. Loango (*Ploetz*), Togoland (*Karsch*).

Ploetz in his catalogue of the species of *Apaustus*, given in the Stett. Ent. Zeit. 1884, places *A. olaus* immediately before his *leucopyga*, which is strictly congeneric with the species described under the same name by Mabille, and made the type of the genus *Acleros*. A good copy of Ploetz's drawing, pl. 744, shows that in form and pattern of marking *olaus* is indeed very near to *leuco-pyga*, Ploetz (*ploetzi*, Mabille); the main difference being the spots in the primaries noted by Ploetz in his original description. An examination of the figure of Ploetz makes it plain, furthermore, that the type was a female. It seems do me quite possible that the insect described was a female of the species previously described by Trimen as *Pamphila* (?) *mackenii*. Karsch apparently is not sure of his identification of this species as given in his article in the Berl. Ent. Zeit. quoted in the synonymy above.

114. A. INSTABILIS, Mab.

Acleros instabilis, Mab. Bull. Soc. Ent. France, (6) vol. ix. p. clxviii (1889).

Hab. Zanzibar (*Mabille*).

There are two females in the collection of Dr. Staudinger,

which are labelled as from Loko, which are plainly referable to this species, which is doubtfully distinct from *A. ploetzi*, Mab. (*leucopygus*, Ploetz). The specimens are smaller in size than is usual in the case of the female of *A. ploetzi*, Mab., and the outer angle of the primaries on the lower side is lighter. The white spots on intervals two and three in the primaries are very large and distinct, more so than in females of *A. ploetzi*, observed by me. Still this may be only a local variety of *A. ploetzi*.

GORGYRA, gen. nov. ·

Antennæ long, slender; club small, gradually thickened, tapering to a fine point; terminal portion bent, but not hooked. Palpi: first joint short; second joint long, profusely clothed with hair, erect, and rising almost or quite to the vertex; third joint long, subconical, porrect, clothed with fine closely appressed hairs. Fore wing: inner margin a little longer than the outer margin; cell about one-half the length of the costa; vein 12 reaching the costa before the end of the cell, veins 7 and 8 from before the end of the cell; the upper and middle discocellulars form an obtuse angle at the end of the cell pointing inwardly, the middle and lower discocellulars form an angle with the apex pointing out-

Head and neuration of *Gorgyra aburæ*, Ploetz, ♂. ⅐.

wardly; vein 5 is nearer vein 4 than 6; vein 3 well before the end of the cell; vein 2 twice as far from the end of the cell as from the base of the wing. Hind wing: the outer margin is evenly rounded and slightly excavated before vein 1 *b*; cell not quite reaching the middle of the wing; vein 7 well before the end of the cell, twice as far from 8 as from 6; discocellulars faint, nearly erect; vein 5 wanting or but faintly indicated; vein 3 just before the end of the cell; vein 2 beyond the middle of the cell; veins 1 *a* and 1 *b* curved; vein 1 *b* clothed on either side with a bundle of long hair-like scales; hind tibiæ almost naked and with two pairs of spurs.

Type *G. aburæ*, Ploetz.

115. G. ABURÆ, Ploetz.

Apaustus aburæ, Ploetz, S. E. Z. vol. xl. p. 359 (1879), vol. xlv. p. 153 (1884).

Hab. Tropical West Africa.

G. DIVERSATA, var. nov.

This form differs from typical *G. aburæ*, Ploetz, in being prevalently lighter in colour on the underside of the primaries and the disc of the secondaries, the darker outer third of the secondaries remaining as in the typical form, and giving the appearance, therefore, of a dark diffuse hind marginal border to the wing.

This form is quite common. About one-half of the specimens collected for me in the Valley of the Ogové belong to it, but I cannot lead myself to believe that it represents a species. Save in the colour modification noted, the specimens otherwise agree absolutely with *G. aburæ*, and there are a number of intergrading forms. (*See next species.*)

116. G. HETEROCHRUS, Mab.

Pamphila heterochrus, Mab. Ann. Soc. Ent. France, (6) vol. x. p. 31, pl. iii. fig. 7 (1890); Novit. Lepidopt. p. 116, pl. xvi. fig. 4 (1893).

Gastrochæta diversata, Mab. MS., in coll. Staudinger.

Hab. Tropical West Africa.

The figure of *G. heterochrus* in the 'Novitates,' was drawn from a specimen in the Staudinger collection, which has been labelled *Gastrochæta diversata* by Mons. Mabille. Another specimen which does not at all agree with the figure in the 'Novitates,' and the duplicate of which was pronounced by Mons. Mabille himself to be a hitherto undescribed species, is labelled in the Staudinger collection as the type of *G. heterochrus.* There has plainly been a misplacement of the labels. I have therefore taken the liberty of applying the name proposed by Mons. Mabille to this new form, of which there are numerous examples in my collection, and which is plainly a mere colour variation of *G. aburæ*, Ploetz (*vide supra*).

117. G. JOHNSTONI, Butl. (Plate II. fig. 6.)

Aeromachus (?) *johnstoni*, Butl. P. Z. S. 1893, p. 673.

Hab. British Central Africa (*Butl.*); French Congo (*Good*).

I have several specimens of this little species from the Valley of the Ogové, agreeing absolutely with the type. It is closely allied to the species described herein as *G. minima*, Holl., but may be distinguished at once by its somewhat larger size, and the fact that the anal extremity of the abdomen is white, which is not the case in *G. minima*, Holl.

118. G. SUBFACATUS, Mab. (Plate II. fig. 11.)

Cobalus subfacatus, Mab. Bull. Soc. Ent. France, (6) vol. ix. p. clxviii (1889).

Hab. Sierra Leone (*Mabille*).

This little species is not white at the end of the abdomen, nor has it the interrupted white line along the inner margin of the secondaries which is conspicuous in *G. aburæ*, Ploetz. The lower side in the type, which is before me, is more prevalently tawny on the costa and at the apex of the primaries, as well as on the disk of the secondaries. Otherwise it closely approximates *G. aburæ*, Ploetz, var. *diversata*, Holl.

119. G. MINIMA, sp. nov. (Plate IV. fig. 24.)

♂. Primaries and secondaries on the upperside black. The primaries are ornamented by two minute spots near the end of the cell, of which the lower one is the larger. Immediately below this spot, in interval 2, is a moderately large sublunate transparent spot, and beyond this in the same series, in intervals 3 and 4, a small spot in each interval. Beyond the cell there is a minute subapical spot. The secondaries have a very small and obscure, scarcely visible, translucent spot at the end of the cell. The primaries and secondaries on the underside are blackish, with The the inner margin of the primaries slightly laved with fulvous. secondaries are obscurely marked with purplish hoary scales. The cilia, both on the upper and lower side, are pale yellowish fuscous. The palpi are black on the upperside, yellowish underneath. The thorax and abdomen on the lower side are blackish.

Expanse 19–20 mm.

Hab. French Congo (*Mocquerys*).

This small species is allied to *G. subfacatus*, Mab., but appears to be quite distinct.

120. G. MOCQUERYSII, sp. nov. (Plate V. fig 10.)

♂. The upperside of the body, the primaries, and the secondaries are black. The primaries are ornamented with three minute subapical spots in the usual position. In some specimens these spots have a tendency to become obsolete. There are two minute white translucent spots at the end of the cell in the primaries, and just below them in interval 2 a subquadrate spot. On vein 1, near the middle in interval 1, is a small subtriangular spot, in interval 3, beyond the end of the cell, a moderately large subquadrate spot. In the male on the secondaries there is a large translucent spot at the end of the cell, and two similar elongated spots beyond the end of the cell on either side of vein 3 at its origin. On the underside the primaries are greenish ochraceous, with the inner half of the wing broadly laved with blackish, shading into fuscous at the outer angle. There is a series of marginal black spots near the apex, and the translucent subapical spots are defined outwardly by blackish markings. There is a fine marginal black line. The cilia are blackish, checkered with whitish on the intervals. On the upperside the cilia are whitish, checkered with blackish at the ends of the nervules. The secondaries on the underside are greenish ochraceous, with the anal angle broadly marked with fuscous. There are three distinct black subcostal spots, a series of black marginal markings, and the translucent spots are narrowly defined by fine blackish lines.

♀. The female is like the male, but lacks the translucent spot at the end of cell of the secondaries.

Expanse ♂ ♀ 25-27 mm.

Types in coll. Staudinger.

Hab. French Congo (*Mocquerys*).

This species is very closely allied to *G. heterochrus*, Mab., from which, however, it may be easily distinguished by the markings of the cilia, and the absence of the patch of light colour which prevails in the secondaries at the anal angle of that species, and by the fact that the lower side of the abdomen is not bright yellowish as in *G. heterochrus*, but greenish ochraceous. There are other distinguishing markings, but these points will suffice at once to separate these species.

121. G. SUBFLAVIDUS, Mab. MS., sp. nov.   (Plate V. fig. 16.)

Pamphila subflavidus, Mab. MS., in Staud. coll.

♂. Primaries and secondaries on the upperside blackish ; cilia of secondaries narrowly white. The primaries are ornamented by a small roundish subapical translucent spot just below the end of the cell, by a small subquadrate spot of the same character in the cell near the lower angle, and by three larger spots on intervals 1, 2, and 3. The spot on interval 1 is subtriangular, on interval 2 subquadrate, and on interval 3 sublunate. The secondaries are ornamented by two translucent wedge-shaped spots on either side of vein 3 near its origin, the uppermost spot being produced beyond the lower. On the lower side the primaries are black, with a small white ray at the base, and with the apical extremity marked with greenish ochraceous. There is a fine marginal black line, two minute blackish spots near the apex, and on either side of vein 5, near the outer margin, whitish markings. The secondaries on the underside are pale straw-colour, with the outer margin and the costa clouded with darker brown markings. On interval 2 there is a dark brownish spot about halfway from the base. A small black spot is found below the costa near the origin of the subcostal nervures, and there is a similar small black spot near the end of the cell. The palpi on the upperside are black, on the lower side straw-yellow, as is also the entire lower side of the thorax and the abdomen. The abdomen towards its

anal extremity is annulated on the lower side with brown, and at
the anal extremity there is a tuft of blackish hairs.

Expanse 28 mm.

Type in coll. Staudinger.

Hab. Usagara, East Africa.

This species is somewhat allied in its markings to *G. aretina*,
Hew., from which, however, it is abundantly distinct. It is
undoubtedly a good species.

122. G. ARETINA, Hew.

Ceratrichia aretina, Hew. Ann. & Mag. Nat. Hist. (5) vol. i.
p. 343 (1878).

Apaustus dolus, Ploetz, S. E. Z. vol. xl. p. 358 (1879), vol. xlv.
p. 151 (1884); Karsch, Berl. Ent. Zeit. 1893, p. 260 pl. vi.
fig. 13.

Gastrochæta albiventris, Mab. MS., in Staudinger coll.

Hab. Old Calabar (*Hew.*); Gaboon (*Good*); Togoland (*Karsch*);
Loko (*Staudinger*).

I have compared the specimens in my collection with the type
of *C. aretina*, Hew., and find them to be identical. The representation of *Apaustus dolus*, Ploetz, given by Karsch, is a most
excellent representation of *G. aretina*, as is shown both by comparison with the insect and with a carefully executed figure of
the type made for me by Mr. Horace Knight, of London. Mons.
Mabille identified the specimens I took with me to Paris as his
Gastrochæta albiventris, comparing them with the type so labelled
in the Staudinger collection, which is now again before me as I
write. I cannot find any record of the publication of this name by
Mons. Mabille, but it may possibly have eluded the vigilance of
the compilers of the ' Zoological Record ' and others engaged in
similar work.

123. G. INDUSIATA, Mab.

Hypoleucis indusiata, Mab. C. R. Soc. Ent. Belg. vol. xxxv.
p. cxiii (1891); Novit. Lepidopt. p. 117, pl. xvi. fig. 6 (1893).

Hab. Cameroons.

This insect is not congeneric with the type of *Hypoleucis*, which
is at best a very doubtful genus. It appears to be more correctly
referred to the genus *Gorgyra*. With the exception of the type
and a single specimen contained in my collection I do not know
of any others in the museums of the world up to the present time.
The type is in the Staudinger collection.

124. G. RUBESCENS, sp. nov. (Plate IV. figs. 17 ♂, 18 ♀.)

♂. Antennæ black, marked with white below before the end of
the club. Palpi black on the upperside, pale yellow beneath.
Upperside of thorax and abdomen dark brown; lower side of thorax
and abdomen obscure ochraceous. The primaries on the upperside
are bright rufous, with the costa and the outer margin broadly

3*

black. There are two translucent spots at the end of the cell, the upper small, the lower linear, fused with each other. There are two translucent wedge-shaped spots on intervals 2 and 3 on either side of vein 3 at its origin, and there are three translucent subapical spots in the usual position, the lower one the largest and elongated, the two upper ones inclined to obsolescence. These translucent spots are only visible when the specimens are held up to the light. The secondaries are bright rufous, like the primaries, with the costa very broadly, and the outer margin more narrowly bordered with black. A long black ray runs from the base to the outer margin before the anal angle. There is a wedge-shaped translucent spot at the end of the cell near its lower edge, and two similar spots on either side of vein 3 at its origin. These spots, like those in the primaries, are only visible when the specimen is held up to the light. On the underside the primaries are dull reddish fuscous, with a pale yellow suffused spot on the inner margin about the middle. A black elongated spot extends from the base outwardly on the cell as far as the inner margin of the translucent spots. These spots are defined outwardly beyond the cell by broad black markings. Near the apex, on the intercostal interspaces, there is a series of submarginal fuscous markings, and the margin is defined by a fine marginal line. The cilia are fuscous. On the underside of the secondaries the prevalent colour is fuscous ochraceous, the translucent spots being distinctly defined on this side, and having a reddish waxy colour. There is a curved series of black submarginal markings extending round the wing, the spots below the costal margin being most conspicuous. There is also a series of small marginal black spots, and a fine black marginal line. The anal angle is touched with dark brown. The black ray running from the base to the outer margin is obscurely indicated on the lower side and interrupted before the anal angle by a blackish annulus, pupilled with pale yellow.

♀. The antennæ, palpi, and body are marked as in the male, but the underside of the body is paler, the lower side of the palpi and the end of the abdomen on the underside being very pale straw-yellow. The primaries on the upperside are black, clothed with greenish scales at the base, along the costa, and the inner margin. The translucent spots in the primaries are bright yellow, standing out conspicuously upon the black ground-colour. The secondaries are marked as in the male, but the black border of the costa is broader and blacker, and the light portions of the wing are bright straw-yellow instead of rufous. The cilia on the upperside at the inner angle both of the primaries and secondaries are whitish. On the underside the ground-colour is bright yellow-ochraceous, with all the black markings as in the male, but broader and more clearly defined upon the pale ground-colour. The spots on the secondaries, which are prevalently bright yellow-ochraceous, are very sharply defined. The black ray on the secondaries running from the base to the inner angle is replaced by three spots—a fine linear spot near the base, a conspicuous round black spot about the middle, and a geminate black spot near the outer margin, all on interval 1.

Expanse, ♂ 26 mm., ♀ 28 mm.

Hab. Valley of the Ogové (*Good, Mocquerys*).

The very great difference in the coloration of this species from that of other species referred by me to the genus *Gorgyra*, and the dissimilarity between the male and female, analogous to that which is found in the various species contained in the genera *Osmodes* and *Pardaleodes*, have long led me to hesitate in referring this species to the genus in which I have finally placed it. A careful anatomical investigation made with bleached specimens under the microscope has made it plain to me that there is almost no structural difference. The form of the palpi, the antennæ, and the neuration is identical with that of the other species referred to *Gorgyra*. The species constitutes a section of the genus separate from its allies on account of the distinct coloration and the diversity in facies between male and female.

GASTROCHÆTA, Muh. MS., gen. nov.

Antennæ slender, moderately long, reaching beyond the middle of the costa; club moderate, gradually thickened, tapering to a fine point, terminal portion bent, but 'not hooked. Fore wing: in the male produced at apex, in the female somewhat more rounded and broader ; the inner margin a little longer than the outer margin. The cell two-thirds the length of the costa. Vein 12 reaching the costa a little beyond the end of the cell.

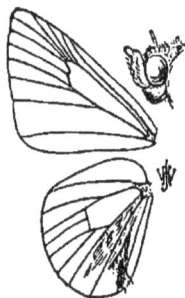

Neuration and palpi of *Gastrochæta mexa*, Hew. ♀.

The upper end of the cell is rounded between veins 11 and 6, and these veins are given forth from this rounded extremity. The upper and middle discocellulars form an obtuse angle with each other pointing inwardly. The middle and lower discocellulars form an obtuse angle with each other pointing outwardly. Vein 5 is slightly nearer vein 4 than vein 6 ; vein 3 from a little before the end of the cell ; vein 2 a little beyond the middle of the cell.

The secondaries are suboval, with the outer margin evenly rounded. The costal and inner margins are straight between the angles. The cell is long, reaching a little beyond the middle of the wing. Vein 7 before the end of the cell, twice as far from 6 as 8; disco-cellulars faint, erect; vein 5 present, equidistant from veins 4 and 6; vein 3 before the end of the cell; vein 2 twice as far from the base as from the end of the cell; veins 1 *a* and 1 *b* straight. Between veins 1 *a* and 1 *b* there is a narrow fold heavily clothed with long tufts of hair-like scales. Interval 1 is likewise clothed heavily with long scales. Palpi: first joint short, second joint long, both heavily clothed with scales; second joint erect, rising to the top of the vertex; third joint short, obtuse, slightly porrect, clothed with fine minute closely appressed hairs.

Type *G. mabillei*, Holl.

Mons. Mabille has designated a number of species by the generic name *Gastrochæta* in his own collection and in the collection of Dr. Staudinger, as well as in my own collection. I discover, however, that he has nowhere published an account of this genus. In the 'Entomological News,' vol. v. p. 28, I published a species under this name as *Gastrochæta mabillei*. As this was the first time that the name appears to have been published, the species to which I have applied it must stand as the type of the genus. In many respects there is a superficial resemblance between the species included in this genus and those included in the genus *Gorgyra*, some of the species of which Mons. Mabille has labelled in the Staudinger collection as belonging to that genus. An examination of the palpi and the neuration, however, instantly reveals the difference.

125. G. MABILLEI, Holl.

Gastrochæta mabillei, Holl. Ent. News, vol. v. p. 28, pl. i. figs. 15, 16 (1894).

Hab. Valley of the Ogové.

126. G. MEZA, Hewitson. (Plate II. fig. 9.)

Hesperia meza, Hew. Ann. & Mag. N. H. (4) vol. xix. p. 79 (1877).

Apaustus batea, Ploetz, S. E. Z. vol. xl. p. 359 (1879), vol. xlv. p. 153 (1884).

Pamphila bubovi, Karsch, Berl. Ent. Zeit. vol. xxxviii. p. 251, pl. vi. fig. 10 (1894).

Gastrochæta varia, Mab. MS., in Staudinger coll.

Hab. Tropical West Africa, from Angola (*Hew.*) to Togoland (*Karsch*). Very abundant at Gaboon.

This species was originally determined for me by Mons. Mabille as *Gastrochæta varia*, Mab., upon comparison with specimens so labelled in his collection and that of Dr. Staudinger, but I cannot find that he has ever published a description under this name.

127. G. OYDEUTES, Holl.

Gastrochæta cybeutes, Holl. Ent. News, vol. v. p. 94, pl. iii.
fig. 15 (1894).

Hab. Valley of the Ogové.

G. OYDEUTES, Holl., var. PALLIDA.

There are two specimens contained in the Staudinger collection
in which the markings on the underside of the secondaries are
quite obscure, and the general coloration of these wings on the
underside is paler. I propose the name *pallida* for this varietal
form.

OXYPALPUS, Wats.

128. O. IGNITA, Mab.   (Plate III. fig. 12.)

♂. *Pamphila ignita,* Mab. Bull. Soc. Ent. France, (5) vol. vii.
p. xl (1877).

Hesperia pyrosa, Ploetz, S. E. Z. vol. xl. p. 356 (1879), vol. xliv.
p. 200 (1883).

♂ (?). *Pamphila gisgon,* Mab. C. R. Soc. Ent. Belg. 1891,
p. clxxii.

♀. *Pamphila gisgon,* Mab. Novit. Lepidopt. p. 95, pl. xiii.
fig. 6 (1893).

Oxypalpus ignita, Watson, P. Z. S. 1893, p. 78.

Hab. Eningo (*Ploetz*); Ogové Valley (*Good*).

Mr. Watson has properly cited *P. gisgon,* Mab., as the female
of *P. ignita,* Mab. All the specimens of *P. ignita* I have ever
seen, some fifty or more, have been males, and all of *P. gisgon* have
been females. I had an opportunity of seeing the type of *P. gisgon,*
and of pointing out to Mons. Mabille that it is a female. In the
'Novitates' he cites it in the plate as of this sex. On the underside *P. ignita* and *P. gisgon* agree very well. There are two forms,
probably seasonal, one smaller and more tawny, the other longer
and darker. Both are represented in my collection and that of
Dr. Staudinger.

129. O. ANNULIFER, Holl.   (Plate III. fig. 11.)

Oxypalpus annulifer, Holl. Ann. & Mag. Nat. Hist., Oct. 1892,
p. 293.

Hab. Valley of the Ogové.

130. O. RUSO, Mab.   (Plate III. fig. 13.)

Pamphila ruso, Mab. C. R. Soc. Ent. Belg. vol. xxv. p. clxxxiii
(1891).

Oxypalpus ruso, Butl. P. Z. S. 1893, p. 669.

Hab. Bagamoyo (*Mabille*); Zomba (*Butler*).

The type I saw in the collection of Mons. Mabille. The species
is not contained either in my own collection or that of Dr. Staudinger. The figure in the plate was drawn from the type.

TEINORHINUS [1], Holl.

Neuration of *Teinorhinus watsoni*, Holl., ♂. ⅞.

131. T. WATSONI, Holl. (Plate III. fig. 10.)

· *T. watsoni*, Holl. Ann. & Mag. Nat. Hist., Oct. 1892, p. 292.
Hab. Gaboon.

OSMODES, Wats.

This is a well-marked genus, the males of which may be distinguished at a glance by the patch of glandular raised scales located on the secondaries near the cell. The females differ greatly from the males upon the side, and in several species seem to be very closely related to each other in the pattern of the markings. In fact it is in many cases possible to discriminate between them only by paying the most careful attention to small points of difference, and by having specimens taken *in coitu*. Fortunately I have been able to satisfactorily solve most of the puzzling problems which the difference of the sexes present, thanks to the possession of vast series of specimens, carefully collected and accompanied by satisfactory observations in the field. It may be said that it seems to me that there is strong probability that several of the species are dimorphic. But further research upon the ground is necessary to establish this supposition.

132. O. LARONIA, Hew. (Plate IV. figs. 1 ♂, 2 ♀.)

O. laronia, Hew. Descript. Hesper. p. 35 (1868).
Plastingia laronia, Ploetz, S. E. Z. vol. xl. p. 356 (1879), vol. xlv.
p. 145–6 (1884).
Osmodes laronia, Wats. P. Z. S. 1893, p. 78.

Hab. Gold Coast, Gaboon.

This species is labelled *Plastingia bicuta* by Mons. Mabille in Dr. Staudinger's collection, but the name has never been published.

133. O. THORA, Ploetz. (Plate IV. figs. 3 ♂, 5 ♀.)

Plastingia thora, Ploetz, S. E. Z. vol. xlv. p. 145 (1884).
Osmodes thora, Wats. P. Z. S. 1893, p. 79.
Hab. Guinea (*Ploetz*), Gaboon (*Good*).

[1] By a misprint in the ' Annals,' originally published as " *Teniorhinus*."

This species is much paler and brighter on the underside than
any other in the genus known to me. It is barely possible that the
species named by me in this paper *Osmodes thops* may be a sensonally
dimorphic form of *thora*. The males agree almost perfectly upon
the upperside, but on the underside *thops* is invariably darker,
and the female of *thops* has the orange spots on the upperside
larger and differing materially in outline.

134. O. ADON, Mab.   (Plate IV. figs. 13 ♂, 15 ♀.)

Pamphila adon, Mab. Bull. Soc. Ent. France, 1889, p. cxlix.

Hab. Sierra Leone, Gaboon.

The description given by Mons. Mabille is based upon a specimen
in which the lower side of the secondaries shows but two silvery
spots. I have a series of about one hundred specimens, which reveal
that there is variation in this respect from specimens which have
no silvery spots at all to those which have five or six. The type
specimen in Mons. Mabille's collection is one which I had the
pleasure myself of communicating to him, and represents a less
spotted form than is quite common. A similar specimen in the
Staudinger collection he has designated as a " type." This species
is undoubtedly dimorphic. I have specimens, larger in size than
the typical form, in which the deep black basal portion of the
primaries is not invaded near the inner margin by a narrow ray of
the bright orange of the median band, as is the case in the type.
But, aside from this, I find no distinction worthy of consideration.

135. O. CHRYSAUGE, Mab.   (Plate IV. fig. 7.)

Pamphila chrysauge, Mab. C. R. Soc. Ent. Belg. 1891, p. clxxii ;
Novit. Lepidopt. p. 93, pl. xiii. fig. 4 (1893).

Hab. Loko (*Mabille*), Cameroons (*Good*).

This species resembles *O. laronia*, Hew., at first sight, the subapical orange spot being confluent with the orange-coloured discal
area of the primaries. But the black marginal band on the
primaries is even on its inward margin and not deeply incised at the
nervules, as is the case in *laronia*. The costal margin of the secondaries is also much more broadly marked with black. Compared with
adosus, a closely allied species, it may be observed that the raised
patch of scales on the secondaries is oval in *chrysauge*, and not so
nearly circular as in *O. adosus*, and is blackish, not reddish, as in the
latter species ; there is a small, linear, velvety mark near this spot
upon the inner margin, which is entirely lacking in *adosus*. Besides
the ground-colour in *O. chrysauge* is slightly paler than in *O. adosus*,
and the black inner marginal border is narrower in the secondaries
than in the last-mentioned species.

136. O. ADOSUS, Mab.   (Plate IV. fig. 10.)

Pamphila adosus, Mab. Bull. Soc. Ent. France, (6) vol. ix.
p. cxlix (1889).

♀. *Pamphila argenteipuncta*, Mab. MS.

Hab. Sierra Leone (*Mabille*); Gaboon (*Good*).

I have the figure of a female *Osmodes* to which Mons. Mabille has affixed the name *argenteigutta*, and to the original type of which in the Staudinger collection he has attached the name *argenteipuncta*. It is undoubtedly the female of the species named *udosus* by him. I know this because I have specimens of the two taken *in coitu*.

137. O. LUX, Holl.   (Plate IV. figs. 23 ♂, 25 ♀.)

Osmodes lux, Holl. Ann. & Mag. Nat. Hist., Oct. 1892, p. 291.

Hab. Valley of the Ogové.

138. O. STAUDINGERI, sp. nov.   (Plate III. fig. 20.)

♀. Antennæ, upperside of head, upper and lower side of thorax, and abdomen dark brown. The palpi on the underside are yellowish. The thorax on the upperside is clothed with a few obscure greenish scales. The primaries and secondaries on the upperside are dark brown. There are two bright yellow confluent spots on the cell near the end, three subapical spots which are situated in the usual place, and a series of spots extending from vein 1 to the subapical spots constituting a sharply defined macular band upon the disc. The lower spot of the series in interval 1 is subtriangular. The spot in interval 2 is elongated, subquadrate, and the largest of the series. The spot in interval 3 is the same form as the spot in interval 2 but smaller. The spots in intervals 4 and 5 are minute, elongated. The lower subapical spot is larger and elongated. The two upper subapical spots are small. In the secondaries there is a small circular yellow spot at the end of the cell, and beyond it an irregularly curved series of five discal spots likewise bright yellow. On the underside the primaries and secondaries are more obscure in colour than on the upperside, the spots and markings being, however, identical in form and position.

Expanse 30 mm.

Hab. Valley of the Ogové.

Type in my collection.

I do not know the male of this species. The solitary female in my collection is, however, so totally distinct from every other species known to me that I do not hesitate to describe it as a new form.

139. O. BANG-HAASII, sp. nov.   (Plate IV. fig. 9.)

♂. Antennæ black. Upperside of palpi, head, thorax, and abdomen rufous-brown. Lower side of the palpi, thorax, and abdomen of the same colour, somewhat more obscure. The primaries on the upperside have the ground-colour bright rufous. The apex, the outer margin, and the outer half of the inner margin are broadly deep black. Beyond the end of the cell there is a broad irregular black spot. The costal margin and the base of the wing as far as the middle of the cell are fulvous, shading outwardly about the middle of the wing into blackish. The secondaries are bright rufous, with the costal margin broadly black, the inner margin somewhat broadly margined with black, the outer margin defined with a moderately broad black marginal line. The cilia are rufous. On the cell is a broad oval patch of raised scales, dark brown in colour. On the underside the wings are more obscurely marked, the spots of the upperside reappearing upon the primaries, but much less sharply defined. The secondaries lack the black costal border and are marked on the disc by a number of minute silvery spots, surrounded by fuscous shadings. Of the spots, the one at the end of the cell is the most conspicuous.

♀. The female presents the usual broad divergence from the male which is characteristic of the genus, and superficially does not apparently differ very widely on the upperside from the female of *O. adosus*, Mab., an allied species. On the underside, however, it agrees almost absolutely with the male in the style of marking.

Expanse, ♂ 26 mm., ♀ 29 mm.

Types in coll. Staudinger.

Hab. French Congo (*Mocquerys*).

This is one of the most distinctly marked species in the genus.

140. O. DISTINCTA, sp. nov.   (Plate IV. fig. 16.)

♂. Very closely allied to *O. chrysauge*, Mab., of which it may be a small variety. It differs from the type of *O. chrysauge* in having the apex more broadly black, the subapical yellow spots not being confluent with the broad orange-yellow discal tract as in *chrysauge*. The outer marginal black border is also relatively wider than in *chrysauge*, and the raised patch of scales on the cell of the secondaries is bright fulvous, not dark brown as in *chrysauge*, elongated, and not broadly oval as in the latter species. On the underside of the secondaries the outer margin is not so broadly marked with fulvous as in *chrysauge*.

Expanse 22 mm.

Hab. Gaboon (*Mocquerys*).

141. O. THOPS, sp. nov.   (Plate IV. figs. 4 ♂, 6 ♀.)

♂. Closely allied to *O. thora*, Ploetz, from which it is to be distinguished by the fact that the black margin of the primaries is narrower than in *thora* and not irregular inwardly as in *thora*, but uniform, and by the fact that the underside of the secondaries is dark brown over the greater portion of the area, whereas in *thora* it is light, the outer margin being pale yellow in *thora*, and the basal half pale glaucous clouded here and there with darker brown.

♀. In the female the spots upon the primaries are broader than in the female of *thora*, while on the secondaries the fulvous spot in *thops* is smaller than the corresponding spot in *thora*.

I have a long series of both males and females, some of the examples taken *in coitu*, and it is perfectly plain that the two species are distinct, though superficially *thops* and *thora* show considerable likeness to each other.

RHABDOMANTIS, gen. nov.

Antennæ: moderately long, nearly two-thirds the length of the costa from the base; club moderate, the terminal portion fine, bent back at right angles. The palpi are as in the genus *Osmodes*.

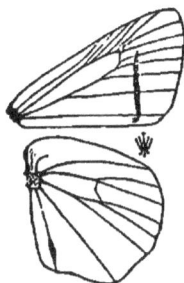

Neuration of *Rhabdomantis galatia*, Hew. ♂.

Primaries: the cell somewhat less than two-thirds the length of the costa; in the male the outer margin is very little less than the inner margin; in the female the outer margin is much less than the inner margin; vein 12 terminating on the costa before the end of the cell; vein 5 nearer 4 than 6; upper discocellular long, outwardly oblique; middle discocellular very short; lower discocellular short; vein 7 arising a little before the upper angle of the cell, vein 2 originating nearly twice as far from vein 3 as vein 3 is from vein 4. In many specimens of the male there is a remarkable sexual brand composed of androconia arranged in a narrow band extending across the disc in almost a straight line from the middle of interval 5 beyond the end of the cell to the inner margin before the outer angle. This is wanting, however, in some specimens, which otherwise are absolutely indistinguishable from the type (*vide infra* var. *sosia*). Secondaries: the cell about half the width of the wing; the discocellulars faint, erect; vein 5 absent; vein 3 originating a little before the end of the cell; vein 2 originating beyond the middle of the cell; vein 1 *b* widely separated from vein 2; vein 1 *a* near its extremity dilated and marked by a distinct sexual brand; vein 7 originating about two-thirds of the distance from the base. The outer margin is evenly rounded as far as vein 2 and much produced at the extremity of vein 1 *b*, then excavated between the extremities of vein 1 *b* and 1 *a*. The female has the neuration like the male, but the wings are longer, relatively narrower, and there is of course an entire absence of the sexual brands or markings. The style of maculation in this sex closely approximates that of the females in the genus *Osmodes*.

Type *R. galatia*, Hew., =*rhabdophora*, Mab.

142. R. GALATIA, Hew.   (Plate III. figs. 8 ♀, 15 ♂.)

Hesperia galatia, Hew. Descript. Hesper. p. 36 (1868).
Pamphila rhabdophorus, Mab. Bull. Soc. Ent. France, (6) vol. ix. p. cxlix (1889).
Dimorphic var. *R. sosia*, Mab.
Pamphila sosia, Mab. C. R. Soc. Ent. Belg. 1891, p. clxxi.

Hab. Old Calabar (*Hewitson*); Gaboon (*Good*); Mozambique (*Mabille*).

I have an enormous series of specimens of this insect, both males and females. It is absolutely impossible to distinguish between the females of *R. galatia* and *R. sosia*. *Sosia* merely differs from *galatia* in being without the raised velvety brand of scales upon the primaries below the end of the cell. Some vestiges of this sexual mark, however, appear in a few specimens. I am perfectly convinced that the insects do not specifically differ from each other, and that we are simply dealing here with dimorphism affecting the sexual stigmata of the male sex. This is a singular fact, and, so far as my observation extends, hitherto unobserved.

PAROSMODES, gen. nov.

Closely allied to the genus *Osmodes*, from which it differs principally in the form of the palpi, the third joint of which is long and porrect, whereas in typical *Osmodes* the third joint is short and suberect.

The antennæ are moderately long, exceeding the middle of the costa. The neuration of the primaries and the secondaries is as in *Osmodes*, and there is likewise at the origin of veins 2 and 3 of the secondaries a raised patch of scales as in *Osmodes*. The primaries, as in the latter genus, have also a long tuft of hairs about the middle of the hind margin; these hairs are ordinarily folded forward against the under surface of the primaries as in *Osmodes*.

Type *P. morantii*, Trim.

143. P. MORANTII, Trim.

Pamphila morantii, Trim. Trans. Ent. Soc. Lond. 1873, p. 122.
Pamphila ranoha, Westw. App. Oates's Matabeleland, p. 353 (1881).
Pamphila morantii, Trim. S. Afr. Butt. vol. iii. p. 311, pl. xii. fig. 3 (1889).
Osmodes ranoha, Butl. P. Z. S. 1893, p. 670.

Hab. South Africa and South Tropical Africa.

144. P. IOTERIA, Mab.

Pamphila icteria, Mab. C. R. Soc. Ent. Belg. vol. xxxv. p. clxxx (1821).
Pamphila zimbaso, Trim. P. Z. S. 1894, p. 74, pl. vi. fig. 17 ♀.

Hab. Manica-land (*Trimen*); Transvaal (*Mabille*).

The type of *icteria* is before me as I write. It is strictly congeneric with *morantii*, Trim.

145. P. HARONA, Westw.

Pamphila harona, Westw. App. Oates's Matabeleland, p. 353 (1881); Trim. P. Z. S. 1894, p. 74.

Hab. Manica-land (*Trimen*); Falls of the Zambezi (*Westwood*).

OSPHANTES, gen. nov.

Antennæ moderately long, slender; club gradually enlarging and terminating in a fine point, the terminal portion being recurved. The palpi are short, appressed, suberect, the first joint short, the second long, both densely covered with thick scales. The third joint is minute, conical. The hind tibiæ are armed with a double pair of spurs. The primaries have the inner margin strongly angulated about the middle and clothed with a long bundle of hairs on the elongated portion of the hind margin, which is as long as the outer margin. Vein 5 nearer 4 than 6. Vein 12 terminating on the costa before the end of the cell. The cell more than half the length of the costa. The secondaries have the neuration as in *Osmodes*. On the lower edge of the cell and about the origin of veins 2 and 3, the cell of the secondaries is naked, marked by an opaque tract, suboval in form, having a glazed appearance. Immediately behind this naked glazed tract is a pocket-like depression on the upperside lying between vein 1 *b* and the lower margin of the cell near the base. The primaries on the underside have the basal portion almost naked toward the base, covered with shining closely appressed scales.

Type *O. ogowena*, Mab.

I was inclined originally to refer this peculiar species to *Osmodes*, to which it is allied, but the very peculiar structure of the hind wing shows such a great divergence from the typical species of *Osmodes* that I feel constrained to erect a new genus for its reception. Furthermore, the coloration of the insect differs in many important particulars from that of typical *Osmodes*. The figure of the insect given in the 'Novitates' by Mabille is sufficiently characteristic, though the spots on the underside are not delineated as they are in the examples before me. They recall somewhat in the specimens I have the maculation of *Padraona zeno*, Trim.

146. O. OGOWENA, Mab.

Plastingia ogowena, Mab. C. R. Soc. Ent. Belg. 1891, p. cxxi; Novit. Lepidopt. p. 94, pl. xiii. fig. 5.

Hab. Valley of the Ogové.

This species was evidently placed by Mons. Mabille with doubt in the genus *Plastingia*, in which he has put a number of other African species. The type of *Plastingia* is *flavescens*, Feld., with which this species has but little in common, save the general style of coloration. It does not agree with any other African species known to me, though coming nearer certain species of *Osmodes* than any others. I have therefore not hesitated to erect a new genus for its reception.

HYPOLEUCIS, Mab.

147. H. TRIPUNCTATA, Mab.

Hypoleucis tripunctata, Mab. C. R. Soc. Ent. Belg. 1891, p. lxix.
Hypoleucis titanota, Karsch, Berl. Ent. Zeit. 1893, p. 254, pl. vi. fig. 5.

Hab. West Africa. Common in the valley of the Ogové.

I have specimens determined by Mons. Mabille and compared with his type, which show that the form figured by Karsch in his excellent plate is identical.

148. H. OPHIUSA, Hew.

Hesperia ophiusa, Hew. Trans. Ent. Soc. Lond. (3) vol. ii. p. 497 (1866); Exot. Lep. vol. v; Hesper. pl. v. figs. 46-48 (1873).
Hypoleucis ophiusa, Mab. C. R. Soc. Ent. Belg. vol. xxxv. p. lxix (1891); Wats. P. Z. S. 1893, p. 82-3; Karsch, Berl. Ent. Zeit. vol. xxxviii. p. 254 (1893).

Hab. Tropical Western Africa.

149. H. CRETACEA, Snell.

Goniloba cretacea, Snellen, Tijd. voor Entom. 1872, p. 27, pl. ii. figs. 4, 5, & 6.
Hesperia camerona, Ploetz, S. E. Z. vol. xl. p. 356 (1879), vol. xliv. p. 48 (1883).
Pamphila leucosoma, Mab. Pet. Nouv. Entom. vol. ii. 1877, p. 114.
Pamphila camerona, Karsch, Berl. Ent. Zeit. p. 250, pl. vi. fig. 9 (1893).

Hab. Tropical West Africa. Common at Gaboon and on Congo ; Togoland (*Karsch*).

The female differs from the male in not having the extremity of the abdomen white and having the wings broader. The figure of *G. cretacea* given by Snellen exaggerates slightly the pale markings on the underside of the secondaries, while that given by Karsch does not show them as they are commonly found. I have specimens, however, which agree nearly with both representations, and which reveal that there is considerable variation in the distinctness of these markings. My collection contains a series of forty specimens taken at different times and places.

150. H. ? ENANTIA, Karsch. (Plate II. fig. 17.)

Hypoleucis enantia, Karsch, Berl. Ent. Zeit. vol. xxxviii. p. 255 (1893).

Hab. Togoland (*Karsch*).

The species was described from a headless example. My conviction is, from the examination of a careful drawing made by Herr Prillwitz, which is reproduced in one of the plates accompanying this article, that we are dealing here with a species of *Ceratrichia* allied to, and perhaps identical with, *C. stellata*, Mab.

CYCLOPIDES, Hübn.

151. C. METIS, Linn.

Papilio metis, Linn. Mus. Lud. Ulr. p. 325 (1764); Syst. Nat. ed. xii. p. 792 (1767); Dru. Ill. Exot. Ent. vol. ii. p. 28, pl. xvi. figs. 3, 4 (1773); Fabr. Syst. Ent. p. 528 (1775); Cram. Pap. Exot. vol. ii. p. 103, pl. clxii. fig. G (1777); Fabr. Spec. Ins. vol. ii. p. 132 (1781); Wulfen, Ins. Capens. p. xxxiii (1786); Fabr. Mant. Ins. vol. ii. p. 85 (1787); Gmel. Syst. Nat. i. 5, p. 2355 (1790); Thunberg, Mus. Nat. Ups. xxiii. p. 9 (1804).
Hesperia metis, Fabr. Ent. Syst. iii. 1, p. 329 (1793); Latr. Enc. Méth. vol. ix. p. 776 (1823).
Cyclopides metis, Hübn. Verz. p. 112 (1816); Trim. Rhop. Afr. Austr. vol. ii. p. 293 (1866); S. Afr. Butt. vol. iii. p. 266 (1889).
Heteropterus metis, Wallgr. Rhop. Caffr. p. 46 (1857); Kirby, Syn. Cat. p. 623 (1871); Auriv. Kongl. Sv. Vet.-Akad. Handl. Bd. xix. no. 5 (1882); Staudgr. Exot. Schmett. vol. i. pl. 100 (1888).
Cyclopides metis, Watson, P. Z. S. 1893, p. 90.
Hab. S. Africa.

152. C. MALGACHA, Boisd.

Steropes malgacha, Boisd. Faune Ent. Madgr. p. 67 (1833).
Hesperia limpopana, Wallgr. K. Sv. Vet.-Akad. Handl. 1857; Lep. Rhop. Caffr. p. 50 (1857).
Cyclopides malgacha, Trim. Rhop. Afr. Austr. vol. ii. p. 294, pl. v. fig. 10 (1866); Grandid. Madgr. vol. xviii. p. 344, pl. lii. figs. 6, 6 a (1887); Trim. S. Afr. Butt. vol. iii. p. 268 (1889); Watson, P. Z. S. 1893, p. 90.
Hab. S. Africa, Madagascar.

153. C. ÆGIPAN, Trim.

Cyclopides ægipan, Trim. Trans. Ent. Soc. Lond. 1868, p. 94, pl. vi. fig. 9; S. Afr. Butt. vol. iii. p. 271 (1889); Watson, P. Z. S. Lond. 1893, p. 90.
Hab. S. Africa.

154. C. WILLEMI, Wallgr.

Heteropterus willemi, Wallgr. K. Sv. Vet.-Akad. Handl. 1857; Lep. Rhop. Caffr. p. 47 (1857).
Cyclopides? willemi, Trim. Rhop. Afr. Austr. vol. ii. p. 296 (1866).
Cyclopides cheles, Hew. Descript. One Hundred New Species

Hesp. ii. p. 42 (1868); Exot. Butt. vol. v. pl. 59. figs..12, 13 (1874).
Cyclopides willemi, Trim. S. Afr. Butt. vol. iii. p. 273 (1889); Watson, P. Z. S. Lond. 1893, p. 90.
Hab. S. Africa, North and South Tropical Africa.

155. C. MENINX, Trim.

Cyclopides meninx, Trim. Trans. Ent. Soc. Lond. 1873, p. 121, pl. i. fig. 12.
Thymelicus meninx, Wallgr. Œfv. K. Vet.-Akad. Forh. 1875, p. 92.
Cyclopides argenteostriatus, Ploetz, S. E. Z. vol. xlvii. p. 110 (1886); Watson, P. Z. S. 1893, p. 90.
Hab. S. Africa.

156. C. SYRINX, Trim.

Cyclopides syrinx, Trim. Trans. Ent. Soc. Lond. 1868, p. 93, pl. v. fig. 8, 1870, p. 387; S. Afr. Butt. vol. iii. p. 269 (1889).
Hab. Cape Colony.

157. C. ABJECTA, Snellen.

Cyclopides abjecta, Snell. Tijd. voor Entom. 1872, p. 52, pl. ii. figs. 15, 16.
Steropes furvus, Mab. Bull. Soc. Ent. France, (6) vol. ix. p. clvi (1889).
Cyclopides uniformis, Karsch, Berl. Ent. Zeit. 1893, p. 245.
Hab. Guinea (*Snellen*); Sierra Leone (*Mab.*); Togoland (*Karsch*).
I think the above synonymy will be found to be correct. The type of Mons. Mabille appears plainly to agree in all particulars with the figure of Snellen, and also with an excellent drawing of *C. uniformis*, Karsch, kindly provided by the author.

158. C. FORMOSUS, Butl.

Heteropterus formosus, Butl. P. Z. S. 1893, p. 670, pl. lx. fig. 8.
Hab. Zomba, British Central Africa.

159. C. QUADRISIGNATUS, Butl.

Cyclopides quadrisignatus, Butl. P. Z. S. 1893, p. 670, pl. lx. fig. 9.
Hab. Zomba, British Central Africa.

160. C. MIDAS, Butl.

Cyclopides midas, Butl. P. Z. S. 1893, p. 671, 1895, p. 265, pl. xv. fig. 6.
Hab. Zomba, British Central Africa (*Butler*).

161. C. LEPELETIERII, Latr.

Hesperia lepeletier, Latr. Enc. Méth. vol. ix. p. 777 (1823).
Cyclopides lepeletierii, Trim. (part.) Rhop. Afr. Austr. vol. ii.
p. 295 (1866); S. Afr. Butt. vol. iii. p. 274 (1889).
Baracus lepeletierii, Watson, P. Z. S 1893, p. 114.

Hab. Southern Africa.

It is with some hesitation that I decline to accept the reference of this and the two following species to Moore's genus *Baracus*, made by Mr. Watson. The thoroughness of Mr. Watson's work should give great weight to his opinions, but in this case, after a careful examination of typical specimens of *C. lepeletierii* and its three congeners, which have been placed in *Baracus*, I am compelled to conclude that the differences are too slight in fact to warrant such a departure from the hitherto received classification of the insects.

162. C. INORNATUS, Trim.

Cyclopides inornatus, Trim. Trans. Ent. Soc. Lond. (3) vol. ii.
p. 179 (1864); Rhop. Afr. Austr. vol. ii. p. 295, pl. v. fig. 11
(1866); S. Afr. Butt. vol. iii. p. 277.
Baracus inornatus, Watson, P. Z. S. 1893, p. 114.

Hab. South Africa.

163. C. ANOMÆUS, Ploetz. (Plate I. fig. 6.)

Apaustus anomæus, Ploetz, S. E. Z. vol. xl. p. 358 (1879),
vol xlv. p. 152.

Hab. Aburi (*Ploetz*).

The type is preserved in the Berlin Museum. A good specimen is contained in the collection of Dr. Staudinger, to which Mons. Mabille has affixed the manuscript name "*acosimus*."

164. C. TSITA, Trim.

Cyclopides tsita, Trim. Trans. Ent. Soc. Lond. 1870, p. 386,
pl. vi. fig. 13; S. Afr. Butt. vol. iii. p. 276.
Baracus tsita, Watson, P. Z. S. 1893, p. 114.
Steropes monochromus, Mab. C. R. Soc. Ent. Belg. 1891, p. lxiv.

Hab. South Africa.

165. C. ARGENTEOGUTTA, Butl.

Cyclopides argenteogutta, Butl. Lepid. Exot. p. 188, pl. lxiv.
fig. 8.

Hab. Nubia (*Butler*).

From the figure given by Dr. Butler it appears a little doubtful whether this species is a true *Cyclopides*.

166. C. (?) PAOLA, Ploetz.

Cyclopides paola, Ploetz, S. E. Z. vol. xlv. pp. 391-2 (1884).

Hab. Angola (*Ploetz*).

I doubt the reference of this species to *Cyclopides*.   The description seems to me to point to a form belonging to some other genus.

167. C. (?) BRUNNEOSTRIGA, Ploetz.

Cyclopides brunneostriga, Ploetz, S. E. Z. vol. xlv. p. 392-3 (1884).

Hab. Pundo Ndongo (*Ploetz*).
This is probably not a true *Cyclopides*.

168. C. ROMI, Robbe.

Cyclopides romi, Robbe, Ann. Soc. Ent. Belg. vol. xxxvi. p. 133 (1892)

Hab. Congo.
I cannot make much out of the brief description of Dr. Robbe. The description would apply perfectly, so far as it goes, to *Cyclopides syrinx*, Trim.

169. C. AMENA, Grose Smith.

Cyclopides amena, H. Grose Smith, Ann. & Mag. N. H. (0) vol. vii. p. 127 (1891).

Hab. Madagascar.
This species is compared by its author to *C. pardalinus*, Butl., which Mr. Watson has referred with its allies to the genus *Ampittia*, but which, after examining the types, I prefer to restore to *Cyclopides*.

170. C. RHADAMA, Boisd.

Steropes rhadama, Boisd. Faune Madgr. p. 69, pl. ix. figs. 10, 11 (1833).
Heteropterus rhadama, Kirby, Syn. Cat. p. 623 (1871).
Cyclopides rhadama, Mab. Grandid. Madgr. vol. xviii. p. 343, pl. lvi. a. figs. 2, 2 a (1887).
Ampittia rhadama, Watson, P. Z. S. 1893, p. 96.

Hab. Madagascar.

171. C. PARDALINA, Butl.

Cyclopides pardalina, Butl. Ann. & Mag. N. H. (5) vol. iv. p. 233 (1879).
Heteropterus pardalinus, Mab. Grandid. Madgr. vol. xviii. p. 345, pl. lii. figs. 7, 7 a (1887).
Ampittia pardalina, Watson, P. Z. S. 1893, p. 96.

Hab. Madagascar.

172. C. MIRZA, Mab.

Cyclopides mirza, Mab. Grandid. Madgr. vol. xviii. p. 342, pl. lii. figs. 3, 3 a (1887).
Ampittia mirza, Watson, P. Z. S. 1893, p. 96.

Hab. Madagascar.

4*

173. C. BERNIERI, Boisd.

Steropes bernieri, Boisd. Faune Madgr. p. 68, pl. ix. fig. 9 (1833).
Cyclopides bernieri, Mab. Grandid. Madgr. vol. xviii. p. 342,
pl. lii. figs. 5, 5 a (1887).
Ampittia bernieri, Watson, P. Z. S. 1893, p. 96.
Hab. Madagascar.

174. C. DISPAR, Mab.

Cyclopides dispar, Mab. Bull. Soc. Ent. France, (5) vol. vii.
p. lxxiii (1877).
Heteropterus dispar, Mab. Grandid. Madgr. vol. xviii. p. 346,
pl. lii. figs. 8, 8 a, 9, 9 a (1887).
Ampittia? dispar, Watson, P. Z. S. 1893, p. 96.
Hab. Madagascar.

175. C. SAOLAVUS, Mab.

Cyclopides saclavus, Mab. C. R. Soc. Ent. Belg. vol. xxxv. p. cvii
(1891).
Hab. Madagascar.

176. C. (?) PHIDYLE, Walker.

Cyclopides phidyle, Walker, the Entomologist, vol. v. p. 56
(1870).
Hab. Hor Tamanib (*Walker*).
I cannot make out this species. I cannot discover where the
type is, if it still exists. The insect remains to be rediscovered.

177. C. (?) LYNX, Moeschler.

Cyclopides lynx, Moeschl. Verhandl. d. k. k. zool.-bot. Ges.
Wien, Bd. xxviii. p. 210 (1879).
Hab. Africa?
Moeschler with some degree of doubt assigns this species to the
African fauna. It may be Asiatic. I do not know it except by
the description referred to above.   .

178. C. (?) STELLATA, Mab.

Ceratrichia stellata, Mab. C. R. Soc. Ent. Belg. 1891, p. lxv;
Butler, P. Z. S. 1893, p. 673.
Cyclopides mineni, Trim. P. Z. S. 1894, p. 72, pl. vi. fig. 16.
Hab. Mombasa (*Mabille*); British Central Africa (*Butler*);
Manica (*Trimen*).
The type is in the collection of Dr. Staudinger. It is a female.
There is also a cotype, a male, which is much smaller and badly
worn, lacking altogether the cilia on the wings and minus the
antennæ. The original reference of this species to the genus
Ceratrichia, which has been followed by Dr. Butler and others, is

not correct, nor is the reference of the species to the genus *Cyclopides* made by Mr. Trimen much better, though certainly more natural than the original location. I have been tempted to erect a new genus for the reception of this and the following form, but with the insufficient material at my command for a close anatomical study I refrain. Manifestly the much shorter antennæ, with obtuse clubs, the long cilia of the primaries and the secondaries, the rounded apex of the primaries, and the different general outline of the wings point to a different generic location than that given by the author of the species.

179. C. (?) PUNCTULATA, Butl.

Ceratrichia punctulata, Butl. P. Z. S. 1895, p. 265, pl. xv. fig. 7.

Hab. British Central Africa (*Butler*).

I think it very doubtful whether this is more than varietally distinct from the foregoing species.

PROSOPALPUS, gen. nov.

Antennæ relatively long, reaching beyond the middle of costa; slender, with a moderately thick and elongated club terminating in a fine point, the terminal portion for a short distance bent, not hooked or recurved. Palpi: first joint short; second joint very long, produced for half of its length beyond the front; both second and third joints heavily clothed with scales; the third joint is long, produced, acute, almost naked. The hind tibiæ have a double pair of spurs. In the primaries the cell is moderately long, its end reaching fully to the middle of the wing; vein 12 terminating slightly before the end of the cell; vein 7 from end of the cell; vein 5 very slightly, if at all, nearer vein 4 than vein 6. The primaries are relatively broad, the outer margin and outer angle evenly rounded. Secondaries: cell short, not reaching to the middle of the wing; vein 5 present, equidistant from veins 4 and 6; vein 7 from before the end of the cell, four times as far from vein 8 as from the end of the cell; vein 8 from very near the base; veins 3 and 4 both from the end of the cell; vein 2 from before the end of the cell; veins 1 *a* and 1 *b* curved; fringes very long; secondaries evenly rounded on the costa and the outer margin to the anal angle; the inner margin nearly straight.

Type *P. duplex*, Mab.

The small species which I have chosen as the type of this genus is very distinct in general appearance from all other species which appear to be in any wise related to it. In the structure of the palpi it approaches somewhat the genera *Gorgyra* and *Parosmodes*. In the form of the wings, broad and evenly rounded, as well as in the almost uniform black coloration, it is widely different from all the species included in those two genera. Instead of being robust, as those species are, it wholly differs, resembling more closely in some respects in the form of its wings the genus *Cyclopides*. It is worthy of remark that the palpi are wanting in the type specimens of *P. duplex* which are contained in the collection of Dr. Staudinger. I have relied for the description of the palpi upon specimens contained in my own collection, which in their remarkable length obscurely suggest the genus *Libythea*.

180. P. DUPLEX, Mab. (Plate III. fig. 17.)

Cobalus duplex, Mab. Bull. Soc. Ent. France, (6) vol. ix. p. clxix (1889).

Hab. Sierra Leone (*Mabille*); Gaboon (*Good*).

181. P. (?) DEBILIS, Ploetz.

Apaustus debilis, Ploetz, S. E. Z. vol. xl. p. 360 (1879), vol. xlv. p. 158 (1884).

Hab. Guinea (*Ploetz*).

I place this species here on the ground of the near relationship of the preceding species to it, as stated by Mons. Mabille.

AMPITTIA, Moore.

182. A. CARIATE, Hew.

Cyclopides cariate, Hew. Descript. One Hundred New Hesperid. p. 44 (1868); Exot. Butt. vol. v. pl. *Cyclopides*, fig. 8 (1874); Mab. Grandid. Madgr. vol. xviii. p. 341, pl. lii. figs. 4, 4 *a* (1887).
Ampittia cariate, Watson, P. Z. S. 1893, p. 96.

Hab. Madagascar.

183. A. COROLLER, Boisd.

Hesperia coroller, Boisd. Faune Ent. Madgr. p. 66, pl. ix. fig. 8 (1833).
Pamphila coroller, Mab. Grandid. Madgr. vol. xviii. p. 364, pl. liv. figs. 1, 1 *a* (1887).
Padraona (?) *coroller*, Wats. P. Z. S. 1893, p. 102.

Hab. Madagascar.

KEDESTES, Wats.

184. K. LEPENULA, Wallgr.

Hesperia lepenula, Wallgr. K. S. Vet.-Akad. Handl. 1857; Lep. Rhop. Caffr. p. 50.
Pamphila? *lepenula*, Trim. Rhop. Afr. Austr. vol. ii. p. 298 (1866).
Cyclopides chersias, Hew. Ann. & Mag. Nat. Hist. (4) vol. xx. p. 327 (1877).
Thymelicus lepenula, Trim. S. Afr. Butt. vol. iii. p. 300, pl. xi. fig. 9 (1889).
Kedestes lepenula, Wats. P. Z. S. 1893, p. 96.

Hab. Southern Africa.

185. K. MACOMO, Trim.

Cyclopides macomo, Trim. Trans. Ent. Soc. Lond. (3) vol. i.
p. 405 (1862)
Pamphila macomo, Trim. Rhop. Afr. Austr. vol. ii. p. 297, pl. vi.
fig. 6 (1866).
Thymelicus macomo, Staud. Exot. Schmett. vol. i. pl. 100, ♀
(♂ *error*) (1888); Trim. S. Afr. Butt. vol. iii. p. 302 (1889).
Kedestes macomo, Wats. P. Z. S. 1893, p. 96.
Hab. Southern Africa.

186. K. CAPENAS, Hew.

Cyclopides capenas, Hew. Descript. One Hundred New Hesperid.
p. 43 (1868); Exot. Butt. vol. v. pl. *Cyclopides*, figs. 2, 3 (1877).
Cyclopides derbice, Hew. Ann. & Mag. Nat. Hist. (4) vol. xx.
p. 327 (1877).
Kedestes capenas, Wats. P. Z. S. 1893, p. 96.
Thymelicus capenas, Trim. P. Z. S. 1894, p. 73.
Hab. Manica.

187. K. CHACA, Trim.

Pyrgus chaca, Trim. Trans. Ent. Soc. Lond. 1873, p. 118, pl. i.
figs. 9, 10; S. Afr. Butt. vol. iii. p. 296 (1889).
Kedestes chaca, Wats. P. Z. S. 1893, p. 96.
Hab. South Africa; South Tropical Africa.

188. K. TUCUSA, Trim.

Pyrgus tucusa, Trim. Trans. Ent. Soc. Lond. 1883, p. 359;
S. Afr. Butt. vol. iii. p. 297 (1889).
Kedestes tucusa, Wats. P. Z. S. 1893, p. 96.
Hab. South Africa.

189. K. MOHOZUTZA, Wallgr.

Hesperia mohozutza, Wallgr. K. Sv. Vet.-Akad. Handl. 1857;
Lep. Rhop. Caffr. p. 50.
Pyrgus mohozutza, Trim. Rhop. Afr. Austr. vol. ii. p. 291, pl. v.
fig. 9 (1866); S. Afr. Butt. vol. iii. p. 294 (1889).
Kedestes mohozutza, Wats. P. Z. S. 1893, p. 96.
Hab. South Africa; South Tropical Africa.

190. K. CALLICLES, Hew.

Cyclopides callicles, Hew. Descript. One Hundred New Hesperid.
p. 42 (1868); Exot. Butt. vol. v. pl. *Cyclopides*, figs. 10, 11
(1877).
Pamphila callicles, Trim. S. Afr. Butt. vol. iii. p. 309 (1889).
Kedestes callicles, Wats. P. Z. S. 1893, p. 96.
Hab. South Africa; South Tropical and North Tropical Africa.

191. K. BARBERÆ, Trim.

Cyclopides barberæ, Trim. Trans. Ent. Soc. Lond. 1873, p. 120,
pl. i. fig. 11 ; S. Afr. Butt. vol. iii. p. 306 (1889).
. *Hab.* Cape Colony ; Mashonaland.

192. K. WALLENGRENII, Trim.

Thymelicus wallengrenii, Trim. Trans. Ent. Soc. Lond. 1883,
p. 361 ; S. Afr. Butt. vol. iii. p. 304, pl. xi. fig. 7 (1889).
Hab. Natal ; Mashonaland.

193. K. NIVEOSTRIGA, Trim.

. *Pamphila?* *niveostriga*, Trim. Trans. Ent. Soc. Lond. (3) vol. ii.
p. 179 (1864) ; Rhop. Afr. Austr. vol. ii. p. 298, pl. vi. fig. 7
(1860) ; Trans. Ent. Soc. Lond. 1870, p. 389.
Thymelicus niveostriga, Trim. S. Afr. Butt. vol. iii. p. 303
(1889).
Hab. S. Africa.

194. K. FENESTRATUS, Butl. (Plate II. fig. 16.)

Baracus fenestratus, Butl. P. Z. S. 1893, p. 673.

Hab. Zomba, British Central Africa.
This species is very closely allied to, if not identical with,
K. wallengrenii, Trim.

195. K. (?) LENTIGINOSA, sp. nov. (Plate IV. fig. 22.)

♀. On the upper surface having the general appearance of a
female of the genus *Osmodes*, to which genus, however, it plainly
cannot be referred, owing to the form of the palpi, which are more
nearly those of the genus *Kedestes*. The palpi, head, thorax, and
abdomen are black. On the underside the palpi are ochraceous,
and the lower side of the abdomen is ochraceous. The primaries
are black, marked with two moderately large subapical yellow spots
in the usual position, two small confluent yellow spots at the end
of the cell, and three moderately large discal yellow spots forming
a diminishing series extending from intervals 1 to 3 below the
cell. The secondaries are crossed beyond the cell on the middle
by a broad curved yellow discal band, diminishing inwardly toward
the base. The primaries have the costal margin and the apex
broadly ochraceous. The cell and the lower half of the wing are
broadly black, upon which the two spots at the end of the cell and
the three forming the discal transverse series on the upperside
reappear sharply defined against the dark ground. The secondaries
are uniformly pale greenish-ochraceous, marked by a few distinct
round black spots, one on the cell near its upper margin between
veins 6 and 7 beyond the end of the cell, one on either side of vein
3 halfway between the cell and the outer margin, one on interval 1
below the cell near the base, a larger one on the same interval
halfway between the base and the outer margin. The cilia of the

primaries brown, on the underside of the secondaries pale ochraceous touched with dark brown near the end of vein 2. Expanse
26 mm.
Type in collection of Dr. Staudinger.
Hab. Gaboon (*Mocquerys*).

ADOPÆA, Billberg.

(*Pelion*, Kirby.)

196. A. THAUMAS, Hufn.

Papilio thaumas, Hufn. Berl. Mag. ii. p. 62 (1776).
♀. *Papilio flavus*, Müll. Prodr. Zool. Dan. p. 115 (1776).
Papilio linea, Wien. Verz. p. 160 (1776).
♀. *Papilio venula*, Hübn. Eur. Schmett. i. figs. 666–669 (1803–
1818).
Thymelicus thaumas, Kirby, Syn. Cat. p. 609 (1871).
Hesperia thaumas, Staud. Cat. d. Lép. p. 35 (1871).
Adopœa thaumas, Billb. Enum. Ins. p. 81 (1820); Wats. P. Z. S.
1893, p. 98.
(For fuller synonymy see works on palæarctic Lepidoptera.)
Hab. North Africa.

197. A. LINEOLA, Ochs.

Papilio lineola, Ochs. Schmett. Eur. i. p. 230 (1808).
Papilio virgula, Hübn. Eur. Schmett. i. figs. 660–663 (1803–
1818).
Thymelicus lineola, Kirby, Syn. Cat. p. 609 (1871).
Hesperia lineola, Staud. Cat. d. Lép. p. 35 (1871).
Adopœa lineola, Wats. P. Z. S. 1893, p. 98.
(For fuller synonymy see works on palæarctic Lepidoptera.)
Hab. Mediterranean coasts of Africa.

198. A. ACTÆON, Esp.

Papilio actæon, Esp. Schmett. vol. i. pl. xxxvi. fig. 4 (1777);
Rott. Naturf. vi. p. 30 (1777).
Papilio actæon, Hübn. Eur. Schmett. i. figs. 488–490 (1798–
1803).
Thymelicus actæon, Kirby, Syn. Cat. p. 609 (1771).
Hesperia actæon, Staud. Cat. d. Lép. p. 35 (1871).
Adopœa actæon, Wats. P. Z. S. 1893, p. 98.
(For fuller synonymy see works on palæarctic Lepidoptera.)
Hab. Mediterranean coasts of Africa.

199. A. HAMZA, Oberth.

Hesperia hamza, Oberth. Étud. Ent. i. p. 28, pl. iii. figs. 2 a, 2 b,
2 c (1876).
Hab. Algeria.

GEGENES, Hübn.

(*Philoödus*, Ramb.)

200. G. NOSTRODAMUS, Fabr.

Hesperia nostrodamus, Fabr. Ent. Syst. iii. 1, p. 328 (1793).
Papilio pygmæus, Cyr. (nec Fabr.) Ent. Neap. pl. li. fig. 5 (1787);
Hübn. Eur. Schnett. i. figs. 458-460 (1798-1803).
Papilio pumilio, Hoffm. Ill. Mag. iii. p. 202 (1804).
Hesperia lefebvrii, Ramb. Cat. Lép. And. p. 90, note (1858).
Pamphila nostrodamus, Kirby, Syn. Cat. p. 598 (1871).
Hesperia nostrodamus, Staud. Cat. d. Lép. p. 35 (1871).
Gegenes nostrodamus, Wats. P. Z. S. 1893, p. 104.
(For full synonymy consult works on European species.)
Hab. Mediterranean coasts of North Africa.

201. G. HOTTENTOTA, Latr.

?♀. *Papilio niso*, Linn. Mus. Ulr. Reg. p. 339 (1764); Syst.
Nat. i. 2, p. 796 (1767).
♂. *Hesperia hottentota*, Latr. Encyc. Méth. vol. ix. p. 777
(1823).
Hesperia letterstedti, Wallgr. K. Sv. Vet.-Akad. Handl. 1857;
Lep. Rhop. Caffr. p. 49.
Pamphila letterstedti, Trim. Rhop. Afr. Austr. vol. ii. p. 300
(1866).
Pamphila hottentota, Staud. Exot. Schmett. vol. i. pl. 99 (1888).
Pamphila hottentota, Trim. S. Afr. Butt. vol. iii. p. 314 (1889).
Gegenes hottentota, Wats. P. Z. S. 1893, p. 104.
♀. *Thymelicus brevicornis*, Ploetz, S. E. Z. vol. xlv. p. 290
(1884).

Hab. Southern and Western Africa as far north as Senegambia.
I follow Mr. Trimen in disregarding the somewhat forcible plea
of Prof. Aurivillius for the identification of Latreille's species with
the *Papilio niso* of Linnæus, and the substitution of the latter
name. The copies of Clerck's figures given by Prof. Aurivillius do
not carry conviction with them. They may apply to several other
obscure African forms as well as to the species named by Latreille,
and the description given by Linnæus is wholly inadequate. We
shall for ever be in the dark as to the species intended by Linnæus.
The identification defended so learnedly by Prof. Aurivillius lacks
altogether that positiveness which in such a case is essential, and
is at best merely opinionative. In letters and orally Mons. Mabille
has stoutly maintained to me the identity of Latreille's species
H. hottentota with the species recently described by Mr. Trimen
under the name *obumbrata* (see p. 59). The females of *G. obumbrata*
are positively undistinguishable from the females of *G. hottentota*, and
I am inclined to think that the form characterized by Mr. Trimen
is a dimorphic variety. Typical males of *G. hottentota* and males
of the form *obumbrata* are found in my collection, having been
taken on the same day and in the same locality *in coitu* with

females which are absolutely inseparable from females of
G. hottentota received from Mr. Trimen and taken at the Cape.
It is worthy of note that all specimens of *G. hottentota* taken in
Angola and northward, so far as they have come under my obser-
vation (I have seen several hundreds of specimens from various
localities), are prevalently smaller than specimens from the Cape.

202. G. OBUMBRATA, Trim.

Pamphila obumbrata, Trim. P. Z. S. 1891, p. 103, pl. ix.
fig. 23, ♂.

Hab. Angola, Gaboon, Liberia, and tropical West Coast of
Africa generally.

This species is excessively common about Gaboon, and, as I
have remarked under *G. hottentota*, appears to be a dimorphic form
of that species. Typical *hottentota* occurs in company with it at
the same places, and the females are absolutely indistinguishable.

203. G. ALBIGUTTA, Mab.

Pamphila albigutta, Mab. Grandid. Madgr. vol. xviii. p. 357
pl. liv. figs. 2, 2 *a* (1887).

Hab. Madagascar, Natal (*in coll. Staudinger*).

The specimen labelled *P. albigutta* by Mabille in the Staudinger
collection is from Natal. It is badly rubbed and worn, but shows
likeness to my *subochracea* (see p. 56). It is doubtfully the insect
figured in Grandidier's ' Madagascar.'

204. G.(?) GAMBICA, Mab.

Pamphila gambica, Mab. Pet. Nouv. Ent. vol. ii. p. 233 (1878).

Hab. Senegambia.

I place this species here without any knowledge of it other than
that derived from the description, in which the author states that
it is very near *G. hottentota*, Latr.

205. G. (?) OCCULTA, Trim.

Pamphila occulta, Trim. P. Z. S. 1891, p. 103.

Hab. South-western Africa, Transvaal.

I place this species here provisionally. Mr. Trimen states that
it is allied in some respects to *G. hottentota*, but fails to describe
the antennæ and palpi, without a knowledge of which the generic
location must be temporarily doubtful. It may turn out to be a
Parnara or a *Baoris*.

PADRAONA, Moore.

206. P. ZENO, Trim.   (Plate III. fig. 6.)

Pamphila zeno, Trim. Trans. Ent. Soc. Lond. (3) vol. ii. p. 179
(1864); Rhop. Afr. Austr. vol. ii. p. 301 (1866); S. Afr. Butt.
vol. iii. p. 313, pl. xii. fig. 2 (1889).

Pamphila splendens, Mab. Pet. Nouv. Ent. vol. ii. p. 114 (1877).

Padraona watsoni, Butl. P. Z. S. 1893, p. 671.

Hab. South Africa, British Central Africa, Somaliland (*in. coll. Holland*).

I have in my possession most beautifully executed drawings of the male and female of the insect recently described by my valued friend Mr. Butler as *Padraona watsoni,* but I am utterly unable to detect any differences of specific value between this form and typical specimens of *P. zeno* which I have received from Mr. Trimen.

A specimen of *Pamphila splendens,* Mab., so labelled by the late Mr. Hewitson, which is found in Dr. Standinger's collection, confirms the view I had reached by the study of Mabille's description that it is the same as *P. zeno,* Trimen.

207. P. (?) COLATTUS, Ploetz.

Apaustus collatus, Ploetz, Berl. Ent. Zeit. vol. xxix. p. 229.

Hab. Delagoa.

This species is known to me only by the copy of the figure of Ploetz, which I have been permitted to examine through the courtesy of Mons. Mabille. Judging from this representation, it is a not distant ally of *P. zeno,* Trimen, differing principally in the narrower fulvous markings of the upperside, and the darker colour of the underside of the wings, which in the drawing are quite black except at the base of the wings. The fulvous spots stand out in bold contrast upon this dark ground.

CHAPRA, Moore.

208. C. MATHIAS, Fabr.

Hesperia mathias, Fabr. Ent. Syst. Suppl. p. 433 (1798); Latr. Enc. Méth. vol. ix. p. 751 (1823).

? *Celænorrhinus thrax,* Hübn. Samml. aussereur. Schmett. (1816–1841).

Hesperia havei, Boisd. Faune Ent. Madgr. p. 64 (1833).

Hesperia insconspicua, Bert. Mem. Acad. Sci. Bologna (1849–50), p. 15, pl. i. figs. 4, 5.

Hesperia thrax, Ledr. Verh. zool.-bot. Ges. Wien, vol. v. pl. iv. figs. 9, 10 (1855).

Hesperia mahopaani, Wallgr. K. Sv. Vet.-Akad. Handl. (1857); Lep. Rhop. Caffr. p. 48 (1857).

Pamphila micipsa, Trim. Trans. Ent. Soc. Lond. (3) vol. i. p. 290 (1862).

Pamphila mohopaani, Trim. Rhop. Afr. Austr. vol. ii. p. 304 (1866).

Epargyreus mathias, Butl. Cat. Fabr. Diurn. Lep. p. 275, pl. iii. fig. 8 (1870).

Pamphila elegans, Mab. Bull. Soc. Zool. France, p. 232 (1877).

Chapra mathias, Moore, Lep. Ceylon, vol. i. p. 169, pl. 70. figs. 1, 1 *a* (1880–81).

?*Pamphila ibara*, Ploetz, S. E. Z. vol. xliv. p. 38 (1883).

Pamphila octo-fenestrata, Snalm. Lep. von Madagascar, p. 108 (1884).

Pamphila mathias, var. *elegans*, Mab. Grandid. Madgr. vol. xviii. p. 356, pl. lv. figs. 4, 4 *a*, 5 (1887).

Pamphila mohopaani, Trim. S. Afr. Butt. vol. iii. p. 324 (1889).

Pamphila inconspicua, Butl. P. Z. S. 1893, p. 672 ; Trim. P. Z. S. 1894, p. 76.

Hab. Africa south of the Sahara, Madagascar, and adjacent islands.

After a very full and thorough study of a great collection of specimens in my possession, coming from all parts of the African continent, including examples from Abyssinia, Zanzibar, the Cape Colony, Angola, Gaboon, and Sierra Leone, and after a diligent comparison with long series before me coming from various parts of continental Asia and the adjacent islands, I am forced to the conclusion, which has already been cautiously maintained by others, that the African insect commonly labelled in collections as *mohopaani*, Wallgr., is identical with the insect named *mathias* by Fabricius. The differences which exist are in most cases merely differences of size, and without locality-labels to show whence the particular specimens come from it would be impossible to distinguish them. The specimens from the region of the Cape are generally a little larger than Indian examples, but 1 have not a few specimens among the three or four hundred examples of the African forms before me as I write which are as small as any I have from India.

Indeed *C. lodra*, Ploetz, which Mons. Mabille maintains, in his correspondence with me, to be simply a small form of *C. mathias*, is smaller than any Indian examples I have in my possession. I do not, however, quite agree with Mons. Mabille in his view, and prefer to still maintain *lodra* in this catalogue as a distinct species (*v. infra*).

209. C. LODRA, Ploetz.

Pamphila lodra, Ploetz, S. E. Z. vol. xl. p. 355 (1879), vol. xliv. p. 45 (1884).

Hab. Tropical West Africa (Gaboon, Cameroons).

This is a diminutive reproduction at first sight of *C. mathias*, Fabr., but while the markings are exactly the same as in that species, it may be easily and invariably separated by attending to the fact not only that it is so small, but that the fringes are pure white, and the undersides of both the primaries and secondaries are dark hoary greyish brown. It may be that this form is, as has been suggested, a mere variety or local race of *C. mathias*, but until we know more about the facts I hesitate to sink the name of Ploetz as a synonym.

210. C. SINNIS, Mab.

Pamphila sinnis, Mab. Pet. Nouv. Ent. vol. ii. p. 285 (1878).
Hesperia weymeri, Saalm. Lep. von Madagascar, p. 107 (1884).
Pamphila sinnis, Mab. Grandid. Madgr. vol. xviii. p. 361 pl. lv.
figs. 1, 2, 2 *a* (1887).
Pamphila albirostris, Grand. Madgr. vol. xviii. p. 361 (1887),
pl. lvi. *a*. figs. 4, 4 *a* (plate not published at date of June 1st, 1895).
Hab. Madagascar.
I have the type of *P. albirostris* before me: it is the male of
C. sinnis, Mab. The characteristic sexual brand on the primaries
shows that the insect is a true *Chapra*.

211. C. WAMBO, Ploetz.

Hesperia wambo, Ploetz, S. E. Z. vol. xlvii. p. 97 (1886).
Hab. Africa (*Ploetz*).
From the description this species would appear to be closely
allied to *mathias*, Fabr. The description is not definite enough to
base any very exact conclusions upon it.

PARNARA, Moore.

I have brought together into this genus an assemblage of species
which are very closely related structurally, and seem to me to be
more properly assigned to *Parnara* than to any other existing
genus. At the same time, it is proper to observe that this
arrangement is in some respects merely tentative. In several
cases the species depart somewhat widely from the type, yet I am
not prepared on this account to separate them, and set up new
genera for their reception.

212. P. BORBONICA, Boisd.

Hesperia borbonica, Boisd. Faune Ent. Madgr. p. 65, pl. ix.
figs. 5, 6 (1833).
Hesperia fatuellus, Wallgr. (nec Hopff.) K. Sv. Vet.-Aknd.
Handl. 1857 ; Lep. Rhop. Caffr. p. 48 (1857).
Pamphila borbonica, Trim. (part) Rhop. Afr. Austr. vol. ii.
p. 303 (1866) ; Mabille, Grandid. Madgr. vol. xviii. p. 360, pl. lv.
figs. 6, 6 *a* (1887) ; Trim. S. Afr. Butt. vol. iii. p. 322 (1889).
Hab. South Africa, Tropical Africa, both East and West, and
the adjacent islands.
This species is very common at Gaboon and at Cameroons.

213. P. GEMELLA, Mab.

Pamphila gemella, Mab. C. R. Soc. Ent. Belg. vol. xxviii.
p. clxxxvii (1884).
Hesperia ellipsis, Saalm. Lepidopt. von Madagascar, p. 109
(1884).
Pamphila gemella, Mab. Grandid. Madagascar, vol. xviii. p. 359.
Gegenes gemella, Mab. l. c. pl. lv. figs. 7, 7*a* (1887).
Hab. Madagascar ; Seychelles (*Abbott*).

214. P. POUTIERI, Boisd.

Hesperia poutieri, Boisd. Faune Ent. Madgr. p. 65 (1833).
Pamphila poutieri, Mab. Grandid. Madagascar, vol. xviii. p. 363.
Gegenes poutieri, Mab. l.c. pl. lv. figs. 8, 8 *a*, 9, 9 *a* (1887).
Hab. Madagascar; Seychelles (*Abbott*).

215. P. DETECTA, Trim.

Pamphila detecta, Trim. Trans. Ent. Soc. Lond. vol. xli. p. 141,
pl. viii. fig. 12 (1893).
Pamphila fallatus, Mab. MS.
Hab. Natal (*Trimen*); Cameroons.
I have several specimens of this species which were taken at
Batanga, Cameroons, by the late Dr. A. C. Good. The insect
laballed *Pamphila fallatus* in the Staudinger collection by Mons.
Mabille, of which I can find no published description, is the
same.

216. P. SUBOCHRACEA, sp. nov.   (Plate IV. fig. 11.)

♂. Head, thorax, and abdomen fuscous, clothed with greenish
hairs.   Underside of palpi, thorax, and abdomen pale greenish
ochraceous.   The primaries and secondaries on the upperside are
dark brown, with a slightly purplish lustre toward the outer
margin.   The costa and the inner margin near the base of both
wings are clothed with greenish hairs.   There is a minute elongated
translucent white spot in the cell on its upper margin toward its
extremity.   There are two minute subapical spots beyond the end
of the cell.   There are three discal spots on intervals 2, 3, and 4
below and beyond the cell, the spot on interval 4 being minute,
the spots on intervals 3 and 2 being subhastate, the latter the
largest.   All these spots are translucent.   On the secondaries
beyond the end of the cell are three small subhastate semi-
transparent discal spots, pale in colour.   On the lower side the
primaries are dark brown on the cell and beyond it on the disc on
intervals 2, 3, and 4.   The inner margin is fuscous grey.   The
costa and the apical area are tawny ochraceous.   The secondaries
are uniformly tawny ochraceous, marked with a dark spot at the
end of the cell and a discal series of dark spots accentuating the
outer extremity of the three limbal spots beyond the end of the
cell.   The cilia are pale ochraceous both on the upper and under
side.   All the spots of the upper surface reappear on the lower
side in both wings, but less distinctly defined than on the upper
surface.   Expanse 31 mm.
Hab. Valley of the Ogové.

217. P. MICANS, sp. nov.   (Plate III. fig. 19.)

♂. Head, thorax, and abdomen bright Mars-brown.   Underside
of abdomen pale ochraceous.   The primaries and the secondaries
are bright Mars-brown, with the costal margin of the secondaries
dark brown.   There are two minute subapical spots in the usual

position, and below and beyond the cell two discal spots, subquadrate in form, on either side of vein 3 near its origin. The lower of these spots is the larger. They are both translucent and waxy yellow in colour. There are two small obscure semi-transparent spots of like colour on the secondaries on either side of vein 3 a little beyond its origin. The margin is slightly darker brown than the body of the wing, and the fringes are paler. On the underside the wings are uniformly pale reddish ochraceous, except the inner margin of the primaries, which is darker, inclining to plumbeous. The spots of the upper surface reappear on the lower side, but far less distinctly defined. Expanse 30 mm.

Hab. Valley of the Ogové.

This very distinct species is represented in my collection by a single specimen.

218. P. (?) URSULA, sp. nov. (Plate II. fig. 4.)

♂. This insect is obscurely brown all over, without any spots or markings whatever.

♀. The female is coloured like the male, but has three elongated subapical spots in the usual position, and two obscure translucent spots on either side of vein 3 of the primaries a little before its origin. The spots are subquadrate.

Expanse, ♂ 26 mm., ♀ 30 mm.

Hab. East Africa.

The types of the males are found in my collection and in the collection of Dr. Staudinger. The only female I have ever seen is contained in the collection of Dr. Staudinger. I refer this insect with some measure of doubt to the genus *Parnara*, with which it in the main agrees in neuration as far as I have been able to ascertain. The insect, however, is not so robust as the other species referred to this genus. The primaries are more rounded on the outer margin and the secondaries somewhat more excavated before the anal angle, in the case of the female. I cannot, however, with the material before me, venture to separate this species from the genus *Parnara*.

SEMALEA, gen. nov.

Antennæ and palpi as in the genus *Baoris*. Primaries: cell about half the length of the wing; vein 5 much nearer 4 than 6; vein 12 terminating on the costa before the end of the cell; vein 7 slightly before the end of the cell; vein 2 one-third of the distance from the base; vein 3 a little before the end of the cell. Secondaries: cell short; vein 5 wanting; discocellulars faint, erect; vein 7 originating well before the end of the cell; vein 3 a little before the end of the cell; vein 2 originating beyond the middle of the cell; the outer margin evenly rounded; the costa slightly produced before the base. The two species referred to this genus are characterized by peculiar sexual markings. In the case of the male of *S. pulvina*, Ploetz, the type of the genus, there is a broad patch of long silky hairs upon the upper surface of the secondaries at the end of the cell, almost entirely covering the cell and the origin of the median nervules. In addition, on the under-side of the primaries there is a broad patch of modified scales, and the inner margin has a fringe of long hairs, which, ordinarily,

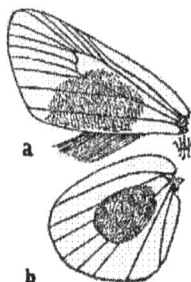

Neuration of *Semalea pulvina*, Ploetz. ⅓.
a. Underside of primary ; *b.* Upperside of secondary.

are folded back upon the under surface of the primaries. In *S. nox*, Mab., the patch of scales on the upper surface of the secondaries is wanting, but upon the primaries on the upper surface there is a broad oval sexual band at the origin of vein 2 below the cell.

Type ♂ *pulvina*, Ploetz.

I have brought these two species together, because of the absolute identity of their neuration and the structure of their palpi and antennæ, and in spite of the wide divergence in the sexual stigmata. At first glance, without a microscopic examination, the two insects look wonderfully alike. There is, however, a remark-able divergence in the sexual stigmata as indicated above. I am, however, more and more inclined to the belief that sexual stigmata cannot be always accepted as the basis of generic subdivisions, in which opinion I know I differ from some authors.

219. S. PULVINA, Ploetz.   (Plate II. fig. 14.)

Hesperia pulvina, Ploetz, S. E. Z. vol. xl. p. 353 (1879).
Trichosemeia pulvina, Wats. P. Z. S. 1893, p. 53.
Cobalus carbo, Mab. Bull. Soc. Ent. France, (6) vol. ix. p. clxix (1889).

Hab. Aburi (*Ploetz*) ; Gaboon (*Good*) ; Sierra Leone (*Mabille*).

I do not agree with Lieut. Watson in placing this species in my genus *Trichosemeia*. The broad patch of velvety scales upon the upper surface of the secondaries is the principal point of resem-blance between this species and the type of the genus. In the form of the wings and the antennæ and the structure of the legs it differs. The female is without the velvety area on the secondaries, and, I strongly suspect, is the insect described by M. Mabille as *Cobalus atrio* (*cf.* genus *Cobalus*). A figure of *C. atrio*, lent me by the author, heightens the probability of this supposition, but without the type before me I will not attempt to express a positive opinion. The insect in the Staudinger collection labelled *Cobalus carbo* in the handwriting of Mons. Mabille is a normal specimen of *S. pulvina*, Ploetz, ♂.

220. S. NOX, Mab. (Plate IV. fig. 20.)

Pamphila nox, Mab. C. R. Soc. Ent. Belg. 1891, p. clxviii.

Hab. Lagos (*Mabille*); Gaboon (*Good*).

This species is apparently very abundant in the valley of the Ogové. I have a large series of specimens.

BAORIS, Moore.

221. B. FATUELLUS, Hopff.

Pamphila fatuellus, Hopff. Monatsber. k. Akad. Wissensch. Berl. 1855, p. 643; Hopff. Peters' Reise n. Mossamb., Ins. p. 417, pl. xxvii. figs. 3, 4 (1862).¹

Pamphila borbonica, Trim. (part.) Rhop. Afr. Austr. vol. ii. pp. 303, 304 (1866).

Hesperia caffraria, Ploetz, S. E. Z. vol. xliv. p. 43 (1883).

Pamphila fatuellus, Trim. S. Afr. Butt. vol. iii. p. 323 (1889).

Pamphila cinerea, Mab. MS.

Hab. Natal, Zanzibar, Gaboon, Cameroons.

This species is very common at Gaboon, and, I think, has generally been confounded with *P. borbonica*, from which, however, it may well be separated, as pointed out by Mr. Trimen. A worn female in the Staudinger collection has been labelled *Pamphila cinerea* by Mons. Mabille.

222. B. MARCHALII, Boisd.

Hesperia marchalii, Boisd. Faune Ent. de Madgr. p. 66 (1833).

Pamphila marchalii, Mab., Grandid. Madagascar, vol. xviii. p. 362, pl. lv. figs. 3, 3 a (1887).

Hab. Madagascar.

Both from the figure and the description I am inclined to think that this species is very near to, if not identical with, *P. fatuellus*, Hopff., in which case Boisduval's name has priority.

223. B. LUGENS, Hopff.

Pamphila lugens, Hopff. Monatsber. k. Akad. Wissensch. Berl. 1855, p. 643; Peters' Reise n. Mossamb., Ins. p. 418, pl. xxvii. figs. 5, 6 (1862); Trim. S. Afr. Butt. vol. iii. p. 318 (1889).

Halpe lugens, Butl. P. Z. S. 1893, p. 673.

Hab. Delagoa Bay, East Tropical Africa.

The genus *Halpe* is not represented in Africa, and Dr. Butler's reference of this species to that genus is in error.

224. B. ILIAS, Ploetz. (Plate V. fig. 17.)

Hesperia ilias, Ploetz, S. E. Z. vol. xl. p. 355 (1879).

Hab. Guinea (*Ploetz*); Gaboon.

What I take to be the *Hesperia ilias* of Ploetz—forming my conclusion from the description of the species given by the author and from a copy of his unpublished drawing of the same—is the insect figured on the Plate. It comes nearer meeting the requirements alike of description and of figure than any other West-African species known to me in nature.

225. B. XYLOS, Mab. (Plate II. fig. 13.)

Pamphila xylos, Mab. Ann. Soc. Ent. France, (6) vol. x. p. 31, pl. iii. fig. 8 (1890).

Hab. Gaboon, Cameroons, Sierra Leone.

Mons. Mabille (*l. c.*) states that he has sufficiently characterized this species in the 'Bulletin' of the preceding year, and contents himself therefore with a figure. By reference to the 'Bulletin' for 1889, I discover that his memory was at fault. He did not describe *P. xylos* in the 'Bulletin' of the year before. Our only knowledge of the species, therefore, must be derived from the figure given in the plate, which, fortunately, is quite recognizable. It represents a damaged male of a species which is quite common on the tropical western coast of Africa. I have a long series of specimens in which, singularly enough, the females are more numerous than the males. The figure given by Mons. Mabille is that of a male minus the abdomen. The female which is represented in the plate does not differ materially in the location and style of the marking from the male, but is generally much larger. I discovered that Mons. Mabille had mingled with this species, in his collection and that of Dr. Staudinger, specimens of the following species, which is abundantly distinct, though presenting a superficial likeness.

226. B. ALBERTI, sp. nov. (Plate II. fig. 21.)

♂. Body and appendages black. Abdomen produced beyond the anal angle of the secondaries. The wings on the upperside are black, with whitish fringes, those of the primaries checkered with black at the ends of the nervules. There are no spots on the secondaries. The primaries are ornamented with three small subapical spots in the usual position, by two large and conspicuous subquadrate spots, one on either side of vein 3 at its origin, the upper one being the smaller of the two. In many specimens there is also a small and faint spot on cell 1, just below the large subquadrate spot on cell 2. On the underside, the wings are marked precisely as on the upperside, save that the inner margin of the primaries is pale, and in some specimens there are traces of an obsolete series of pale submarginal markings on the secondaries.

♀. The female is marked like the male, save that on the under-

5*

side there is a well-defined row of pale submarginal markings on
the primaries, and a less well-defined series of similar markings
on the secondaries. The wings in this sex are broader, more
rounded, and less produced at the apex of the primaries than in
the male, and the abdomen is stouter and shorter than in that sex,
not reaching beyond the anal angle of the secondaries.

Expanse, ♂ 26–29 mm., ♀ 32–34 mm.

Hab. Valley of the Ogové, Cameroons, Sierra Leone.

I name this species in honour of my little friend Albert Good,
the only child of Dr. A. C. Good, one of the heroes of the Dark
Continent, whose death last November, a few days after his return
from a long and trying journey into the interior of the Cameroons,
has filled the hearts of a host of friends and admirers with profound sorrow. " Bertie," though not yet in his teens, is represented in my collections by numerous specimens taken by his own
hands, and is no doubt the youngest entomologist who has thus
far collected amidst the jungles of " Gorilla-land."

227. B. ARELA, Mab. (Plate II. fig. 20.)

Hypoleucis arela, Mab. C. R. Soc. Ent. Belg. 1891, p. lxix.

Hab. Gaboon, Ogové Valley.

This species, for the identification of which in my collection I
am indebted to Mons. Mabille, is quite common about Gaboon.
Mons. Mabille had affixed the name *atimus* to several specimens of
this species in his collection at the time I visited him. They were
undoubtedly *arela.*

228. B. ARGYRODES, Holl.

Parnara argyrodes, Holl. Ent. News, vol. v. p. 93, pl. iii. fig. 11
(1894).

Hab. Valley of the Ogové.

229. B. MABEA, Holl.

Parnara mabea, Holl. Ent. News, vol. v. p. 92, pl. iii. fig. 12
(1894).

Hab. Valley of the Ogové.

230. B. LEUCOPHÆA, Holl.

Parnara leucophæa, Holl. Ent. News, vol. v. p. 93, pl. iii. fig. 14
(1894).

Hab. Valley of the Ogové.

231. B. UNISTRIGA, Holl.

Parnara unistriga, Holl. Ent. News, vol. v. p. 30, pl. i. figs. 13,
14 (1894).

Hab. Valley of the Ogové.

232. B. MELPHIS, Holl.

Parnara melphis, Holl. Ent. News, 1894, p. 31, pl. i. fig. 18.
Hab. Valley of the Ogové.

233. B. MALTHINA, Hew.

Hesperia malthina, Hew. Ann. & Mag. Nat. Hist. (4) vol. xviii.
p. 457 (1876).
Pamphila euryspila, Mab. C. R. Soc. Ent. Belg. vol. xxxv.
p. clxxix (1891).
Proteides euryspila, Mab. Novit. Lepidopt. p. 117, pl. xvi. fig. 5
(1893).
Hab. Sierra Leone (*Mabille*); Gaboon (*Good*).

The only specimen of the female which I have ever seen is
contained in my collection, and was taken at Batanga, Cameroons.
It does not differ materially from the male, save that there is an
additional translucent spot upon the fore wing in cell 1, and the
large white spot on the underside of the secondaries is much larger
than in the male, extending farther outwardly and inwardly.

234. B. STATIRA, Mab.

Pamphila statira, Mab. C. R. Soc. Ent. Belg. vol. xxxv. p. clxxx
(1891); Novit. Lepidopt. p. 114, pl. xvi. fig. 2 (1893).

♂. The type of this species was a female contained in the
collection of Dr. Staudinger. The collection also contains a male,
which differs from the female on the upperside in lacking the spot
in the cell of the primaries, and in having, in the example before
me, the uppermost of the three subapical spots obsolete. In
the secondaries, the spots at the end of the cell which are conspicuous in the female on the underside and faintly appear on the
upperside are also lacking, and the discal spots are somewhat
smaller than in the female.
Hab. French Congo (*Mocquerys*).

235. B. STATIRIDES, sp. nov. (Plate V. fig. 6.)

♀. Palpi on the upperside, head, thorax, and abdomen on the
upperside black, clothed with fuscous scales. Palpi on the underside whitish; thorax and anal extremity of the abdomen pale
fuscous. Primaries black on the upperside, with two widely
separated minute elongated spots near the end of the cell, two
subapical spots in the usual position, and a discal series of four
spots, the lowermost of the series on interval 1 cuneiform, the
next on interval 2 subquadrate, and the largest of the series, and
the two succeeding ones on intervals 3 and 4, subquadrate, the last
smaller than the one preceding it. The secondaries beyond the
cell are adorned with a broad irregularly curved white macular
band, running from before the end of the cell inwardly and
widening to vein 1 b. The primaries on the underside are black,
with the spots as on the upperside. The secondaries are creamy
white, with the outer margin broadly black. There is a con-

spicuous patch of black raised scales situated on interval 1 below
the cell, and extending outwardly on either side of vein 2 at its
origin. Expanse 34 mm.
Type in collection of Dr. Staudinger.
Hab. Valley of the Ogové (*Mocquerys*).

236. B. NETOPHA, Hew.

Hesperia netopha, Hew. Ann. & Mag. Nat. Hist. (5) vol. i. p. 345
(1878).
Hesperia roncilgonis, Ploetz, S. E. Z. vol. xliii. p. 450 (1882).
Pamphila roncilyonis, Trim. Trans. Ent. Soc. Lond. vol. xli.
p. 139, pl. viii. fig. 11 (1893).
Pamphila cojo, Karsch, Berl. Ent. Zeit. vol. xxxviii. p. 250,
pl. vi. fig. 7 (1893).
Var. NYASSÆ, Hew. (Plate I. fig. 8.)
Hesperia nyassæ, Hew. Ann. & Mag. Nat. Hist. (5) vol. i.
p. 345 (1878).
Hab. Natal, Mashonaland, Angola, Gaboon, Cameroons,
Togoland.
This is one of the most singularly coloured species of the group.
I have a good series of specimens from Gaboon and Cameroons,
which agree very well with specimens received from Mr. Trimen,
who obtained them from Mr. F. C. Selous, who took them in
Manica. The type of *Hesperia nyassæ*, Hew., I think is a female.
It is larger and paler on the underside than any specimens I have
seen from other localities. I cannot, however, bring myself to
believe that it is anything more than a variety of *B. netopha*. It
is worthy of note that there is much variation in the ground-colour
of the underside of the wings in this species. No two specimens
in a series of ten or twelve are exactly of the same shade, and the
ground-colour runs from a pale yellowish ochraceous to a pale
reddish brown, tinged with pink. The three small subapical spots
which appear in a majority of specimens are wanting in others.
They are variable.

237. B. MONASI, Trim.

Pamphila monasi, Trim. S. Afr. Butt. vol. iii. p. 317.
Hab. Natal.

238. B. TARACE, Mab.

Pamphila tarace, Mab. C. R. Soc. Ent. Belg. vol. xxxv. p. clxxix
(1891); Novit. Lepidopt. p. 114, pl. xvi. fig. 1 (1893).
Hab. Sierra Leone.

239. B. SUBNOTATA, Holl.

Parnara (?) *subnotata*, Holl. Ent. News, vol. v. p. 94, pl. iii.
fig. 13 (1894).
Pamphila rufipuncta, Mab. MS. in Dr. Staudinger's collection.
Hab. Valley of the Ogové.

240. B. NIVEICORNIS, Ploetz.

Hesperia niveicornis, Ploetz, S. E. Z. vol. xliv. p. 3 (1883).

Hab. Angola.

I only know this species from the figure of Ploetz. It is very remarkably ornamented upon the underside of the wings, and the description given is sufficient to enable its identification.

241. B. (?) NEOBA, Mab.

Pamphila neoba, Mab. C. R. Soc. Ent. Belg. vol. xxxv. p. clxxxviii (1891).

Hab. Cameroons (*Mabille*).

I only know this species from the description and the drawing of the type furnished me by Mons. Mabille. It is impossible from either to be sure of the species or its generic location.

242. B. (?) ZEPHORA, Ploetz.

Apaustus zephora, Ploetz, S. E. Z. vol. xlv. p. 156 (1884).

Hab. Angola (*Ploetz*).

I do not know this species save by the description. It does not seem to apply to any of the species known to me in nature.

243. B. (?) BAUEI, Ploetz.

Hesperia bauri, Ploetz, S. E. Z. vol. xlvii. p. 98 (1886).

Hab. Aburi.

I do not know this species, and locate it here provisionally.

244. B. (?) MURGA, Mab.

Pamphila murga, Mab. Ann. Soc. Ent. France, (6) vol. x. p. 31 (1890).

Hab. Caffraria (*Mabille*).

Mons. Mabille compares this species with *P. natalensis*, Ploetz. I cannot discover that Ploetz ever published a species under the name of *natalensis*. From the description, the insect seems to be possibly a *Baoris*, but it may be a *Pardaleodes*. I locate it here provisionally, as I am unable to learn anything about it from the author of the species.

245. B. (?) HOLTZII, Ploetz.

Hesperia holtzii, Ploetz, S. E. Z. vol. xliv. p. 43–4 (1883).

Hab. Angola (*Ploetz*).

I can make nothing out of either the description or the figure of Ploetz. The insect represented seems both to myself and to Mons. Mabille to be a possible variety of *C. mathias-mohopaani*. I am, however, very strongly inclined to the opinion that it is the same insect recently described by Mr. Trimen under the name *Pamphila monasi* (q. v.).

246. B. (?) AYRESII, Trim.

Pamphila ayresii, Trim. S. Afr. Butt. vol. iii. p. 321, pl. xii. fig. 1 (1889).

Hab. South Africa; South Tropical Africa.

PLATYLESCHES, gen. nov.

Allied to *Parnara*, Moore. The thorax and head are very broad, and the general appearance of the body is more robust than in *Parnara*. The antennae are more than half as long as the costa of the primaries, slender, terminating in a stout club, with a strongly recurved hook at its end. The palpi are broad, flattened horizontally, appressed, heavily clothed with long scales upon the first and second joints, and with the third joint (which is minute, acute, and situated on the outer edge of the horizontally widened second joint) naked. The wings are relatively somewhat narrower than in the genus *Parnara*, with the outer margin of the primaries nearly straight, or, as in *P. picanini*, Holl., slightly excavated above the outer angle. The secondaries are more or less lobed at the anal angle in the male. The neuration of the wings does not materially differ from that in *Parnara*, so far as I have been able to determine with the limited material at my disposal.

Type *P. picanini*, Holland.

247. P. PICANINI, Holl.

Parnara (?) *picanini*, Holl. Ent. News, vol. v. p. 91, pl. iii. fig. 9 (1894).

Pamphila grandiplaga, Mab. MS. in Staudinger collection.

Hab. Valley of the Ogové.

248. P. MORITILI, Wallgr.

Hesperia moritili, Wallgr. K. Sv. Vet.-Akad. Handl. 1857; Lep. Rhop. Caffr. p. 49 (1857).

Pamphila (?) *moritili*, Trim. Rhop. Afr. Austr. vol. ii. p. 305 (1866).

Hesperia neba, Hew. Ann. & Mag. Nat. Hist. (4) vol. xix. p. 84 (1877).

Pamphila moritili, Trim. S. Afr. Butt. vol. iii. p. 319, pl. xii. fig. 4 (1889).

Hab. South Africa; South Tropical Africa.

249. P. GALESA, Hew. (Plate I. fig. 7.)

Pamphila galesa, Hew. Ann. & Mag. Nat. Hist. (4) vol. xix. p. 79 (1877).

Hab. West Africa.

I only know this species from the type, which is preserved in the British Museum. It is a very robust insect, and very closely allied to *H. nigerrima*, Butl.

250. P. NIGERRIMA, Butl.   (Plate II. fig. 12.)

Halpe nigerrima, Butl. P. Z. S. 1893, p. 672.

Hab. British Central Africa.

This species is exceedingly close to *P. galesa*, Hew. The only difference I can detect is in the form of the macular band on the upperside of the secondaries, which is more irregularly curved in *galesa* and has a slightly different direction, and in the presence in *nigerrima* of a narrow white costal streak on the underside of the primaries at the base. This last feature seems to be lacking in *galesa*.

251. P. CHAMÆLEON, Mab.

Pamphila chamæleon, Mab. C. R. Soc. Ent. Belg. vol. xxxv. p. clxxix (1891); Novit. Lepidopt. p. 115, pl. xvi. fig. 3 (1893).

Hab. Sierra Leone.

Mons. Mabille compares this species with his *P. grandiplaga*, which in his letter he identifies as my *P. picanini*. *Grandiplaga* is apparently a MS. name. My learned friend is in the habit of affixing names to specimens coming into his possession, and has given them currency now and then in his papers and through collections which he has labelled, without having published a description of the species. This has led to a great deal of bewilderment on my part in several cases and an inordinate consumption of valuable time in quest of the place in literature where the supposed description, which ought to have been published, might be found. Unpublished names of species should not be referred to, except it be with a distinct statement that they are such.

252. P. AMADHU, Mab.   (Plate V. fig. 11.)

Pamphila amadhu, Mab. C. R. Soc. Ent. Belg. vol. xxxv. p. clxxviii (1891).

Pamphila heterophyla, Mab. l. c.

Baoris? *amadhu*, Butl. P. Z. S. 1893, p. 672.

Hab. Transvaal, Natal (*Mabille*); British Central Africa (*Butler*).

I have before me the types of *P. amadhu* and *P. heterophyla*, belonging to Dr. Staudinger, and am satisfied of the identity of the two forms. The type of *P. heterophyla* is simply a dwarfed and somewhat worn example of *P. amadhu*. The insect is closely allied to *P. moritili*.

253. P. BATANGÆ, Holl.

Parnara batangæ, Holl. Ent. News, vol. v. p. 92, pl. iii. fig. 10 (1894).

Hab. Valley of the Ogové.

254. P. NIGRICANS, sp. nov.   (Plate II. fig. 15.)

♂. Antennæ black, marked with white before the extremity on

the upperside. The uppersides of the head, thorax, and abdomen are black, with the anal extremity of the abdomen white tipped with a tuft of black hairs. The palpi and the pectus on the lower side are white. The lower side of the thorax is grey. The lower side of the abdomen is black annulated with white. The primaries on the upperside are black. The cilia on the primaries are black marked with white at their extremities near the outer angle. The cilia of the secondaries are white, very conspicuously so near the anal angle. The primaries are marked with two subapical spots in the usual position, two elongated minute spots on the cell near its end, one on its upper margin and one on its lower, and by a transverse discal series of four spots, of which the one on interval 1 is minute and subtriangular, situated on vein 1, the spot on interval 2 is subquadrate, excavated externally, and separated from the other spot in the cell by the median nerve. Beyond this spot on intervals 3 and 4 are two smaller spots. The secondaries are crossed about the middle by an irregularly-curved series of five or six white semi-translucent spots. On the underside, the primaries are black, darkest at the base. There is a fine white costal ray near the base. The spots of the upperside reappear, but less distinct than on the upperside, and above and beyond the spot on interval 1 is a white curved ray uniting on its curved upper margin the two lower spots of the discal series. The secondaries are black, most conspicuously so in the region of the anal angle. The inner margin and the outer margin from the outer angle to the extremity of vein 2 are sprinkled with grey scales, and the nerves are likewise clothed with grey scales, causing them to be picked out distinctly upon the dark background. The white discal series of spots reappears on the underside, the terminal spot of the series located on vein 1 b being the most conspicuous, whereas on the upperside it is least conspicuous and appearing as a large triangular white patch.

♀. The female is like the male, but with broader and more rounded wings.

Expanse, ♂ 28 mm., ♀ 30 mm.

Types in coll. Staudinger.

Hab. Freetown (*Preiss*); Gaboon (*Mocquerys*).

The male is labelled in the Staudinger collection *P. leucopyga*, Ploetz, but this determination is wholly in error. *Leucopyga* of Ploetz is an *Acleros* and a wholly different insect. This species is closely related to *P. moritili* and its allies.

KATREUS, Wats.

255. K. JOHNSTONII, Butl. (Plate II. fig. 18.)

Astictopterus johnstonii, Butl. P. Z. S. 1887, p. 573.

Katreus johnstonii, Watson, P. Z. S. 1893, p. 115; Holl. Ent. News, Jan. 1894, pl. i. fig. 8.

Hab. Cameroons, Gaboon.

PARDALEODES, Butl.

256. P. EDIPUS, Cram.

♂. *Papilio edipus*, Cram. Pap. Exot. iv. pl. ccclxvi. figs. E, F (1782).
Pardaleodes edipus, Butl. Ent. Mo. Mag. vol. vii. p. 96 (1870); Kirby, Syn. Cat. p. 625 (1871).
Plastingia edipus, Ploetz, S. E. Z. vol. xl. p. 358 (1879), vol. xlv. p. 148 (1884).
Pardaleodes edipus, Watson, P. Z. S. 1893, p. 117.
♀. *Cyclopides sator*, Doubl. & Hew. Gen. Diurn. Lep. pl. lxxix. fig. 4.
Pamphila? *sator*, Westw. l. c. p. 523 (1852).
Pardaleodes sator, Kirby, Syn. Cat. p. 625 (1871).
Plastingia sator, Ploetz, S. E. Z. vol. xl. p. 358, & vol. xlv. p. 148.
Pardaleodes sator, Watson, P. Z. S. 1893, p. 117.

Hab. Tropical West Africa.

After a very close study of the matter in the light of long series of specimens, consisting of several hundreds of examples, I am satisfied that this is the correct synonymy of this species, which is very closely allied to the next, and with which it has been no doubt, so far as the female of that is concerned, frequently confounded. The crucial test for discriminating between the two species is the fact that in *P. incerta*, Snell., the anterior wings in both sexes show no translucency in the spots above vein 2, whereas in *P. edipus* the spots between veins 2 and 3 and 3 and 4, the spots at the end of the cell, and the three small subapical spots are invariably translucent. By holding the specimens up to the light, it is always possible to decide to which of the two species they belong.

I am at a loss to account for the fact that several authors report the male and the female of both *P. edipus* and *P. sator* to have been contained in collections examined by them. This is done by Ploetz in his paper upon the Lepidoptera collected by Buchholz. So far as my observations extend, every specimen of *P. sator*, correctly determined to be such by comparison with the very good figure given by Doubleday and Hewitson in their work, has been a female. I have seen hundreds of specimens, and many pairs taken *in coitu*, and am sure of this determination.

257. P. INCERTA, Snellen.

♂. *Pamphila incerta*, Snellen, Tijd. voor Entom. 1872, p. 29, pl. 10. figs. 10, 11, 12.
♀. *Hesperia coanza*, Ploetz, S. E. Z. vol. xliv. p. 232 (1883).
Pardaleodes coanza, Watson, P. Z. S. 1893, p. 117.

Hab. Tropical West Africa.

The female of this species resembles the male of the preceding, *P. edipus*, but the point of discrimination enables an easy decision to be made in all cases, as I have already shown.

258. P. HERILUS, Hopff.

Pamphila herilus, Hopff. Monatsber. d. k. Akad. d. Wissensch.
Berl. 1855, p. 643; Peters' Reise n. Mossamb., Ins. p. 419,
pl. xxvii. figs. 7, 8 (1862).

Hab. Querimba, East Africa (*Hopffer*).

Hopffer states that the types of this species were males. From
the figure, I should say that they were females. The figure represents apparently a dwarfed female of *P. edipus*, and closely resembles
such which I have from Gaboon.

259. P. REICHENOWI, Ploetz. (Plate III. fig. 18.)

♀. *Plastingia reichenowi*, Ploetz, S. E. Z. vol. xl. p. 357 (1879),
vol. xlv. p. 147 (1884).

♂. *Pardaleodes festus*, Mab. Ann. Soc. Ent. France, (6) vol. x.
p. 33, pl. iii. fig. 2 (1889).

Hab. Tropical West Africa.

There is not a particle of doubt of the correctness of the above
synonymy. I have specimens taken *in coitu* of the male and female
of this species. The males have been repeatedly determined for
me as *P. festus* by Mons. Mabille, and agree perfectly with the
figure he gives. The females agree with Ploetz's type of *P. reichenowi*, which is preserved at the Berlin Museum, and is represented
in the plate accompanying this paper.

260. P. XANTHOPEPLUS, Holl. (Plate III. figs. 9 ♂, 16 ♀.)

Pardaleodes xanthopeplus, Holl. Ann. & Mag. N. H. (6) vol. x.
p. 289 (1892).

Hab. Valley of the Ogové.

261. P. DULE, sp. nov. (Plate. III. fig. 21, ♂ ♀.)

Allied to *P. reichenowi*, Ploetz = *festus*, Mab.

♂. Primaries deep black, slightly clothed with greenish scales
near the base. The wing is marked with eleven spots as follows :—
two small oval spots at the end of the cell, one above each other,
and above them a minute linear spot; three small oval subapical
spots forming a series curving inwardly; a small round spot in
interval 4; a triangular spot in interval 3; a large subquadrate
spot in interval 2; (these three spots form a transverse series
running inwardly towards the margin). The large spot is followed
in interval 1 by a triangular orange-yellow spot, diminishing
towards the inner margin. There is also an obscure orange-
yellow spot in interval 1 towards the base. All the spots are
translucent, except the two in interval 1, which are opaque.
The secondaries are bright orange-yellow, paler than in *P. festus*,
with the costal margin and outer angle broadly black. The cell
near the base and the inner margin are clothed with fuscous
hairs.

On the underside, the primaries are blackish, the spots of the

upperside reappearing, but pale ochraceous. The costa is, furthermore, laved with pale ochraceous from the base to the region of
the subapical spots, and in interval 5 there is a pale ochraceous
area, in the middle of which there is a minute white dot circled
with blackish. A pale yellowish-grey ray connects the lowermost
spot of the discal series with the outer angle. The secondaries
are pale ochraceous, with the costa on the inner two-thirds marked
with irregular blackish spots. There is a subtriangular blackish
spot near the outer angle, a black spot in interval 1 b near the cell,
and a smaller similar spot surmounted with a V-shaped blackish
mark on the same interval near the anal angle. The innermost of
these last two blackish spots is supplemented on the side of the
base with a small chalky-white spot. There are in addition a
number of obscure transverse brownish lines and obscure submarginal hastate markings.

♀. The female is like the male, except that it wholly lacks the
markings on interval one in the primaries, and the markings on
the underside of the secondaries are not so distinct. The outline of the wings, furthermore, is broader.

Expanse, ♂ ♀ 36 mm.

Hab. Bulé country, Cameroons.

This species may be easily distinguished from *P. reichenowi* by
the deeper black of the primaries, the smaller size of the spots,
and the fact that none of them are confluent, as in *P. reichenowi.*
There is no black border on the inner two-thirds of the secondaries and no yellow spot in the cell of the secondaries, the yellow
of the hind wing running almost to the base. A further distinction is the absence of the checkered fringes of the primaries
on the upper surface. The fringes are slightly checkered on the
underside.

262. P. ASTRAPE, Holl. (Plate IV. fig. 12.)

Pardaleodes astrape, Holl. Ann. & Mag. N. H. (6) vol. x. p. 290
(1892).

Pardaleodes parcus, Karsch, Berl. Ent. Zeit. vol. xxxviii. p. 258
(1893).

Hab. Valley of the Ogové (*Good*) ; Togoland (*Karsch*).

263. P. ARIEL, Mab.

Pamphila ariel, Mab. Pet. Nouv. Entom. vol. ii. p. 285 (1878).

Pardaleodes ariel, Mab. Grandid. Madgr. vol. xviii. p. 340,
pl. liii. figs. 10, 10 a, 11 (1887).

Hab. Madagascar.

264. P. PUSIELLA, Mab.

Pardaleodes pusiella, Mab. Bull. Soc. Zool. France, 1877, p. 237.

Hab. Landana (*Mabille*).

I cannot find out anything about this species.

265. P. LIGORA, Hew.

Hesperia ligora, Hew. Ann. & Mag. N. H. (4) vol. xviii. p. 450 (1876).
Carystus thersander, Mab. Ann. Soc. Ent. France, (6) vol. x. p. 30, pl. iii. fig. 5 (1890).
Carystus? thersander, Holl. Ent. News, vol. v. pl. i. fig. 17 (1894).
Hab. Angola *(Hew.)*; Sierra Leone *(Mab.)*; Cameroons *(Good).*

After a careful examination of the structure of this species, although it greatly exceeds in size any other species of *Pardaleodes* known to me, and the primaries are more pointed than in the type of the genus, I cannot find anything to justify its separation from *Pardaleodes.* With *P. xanthioides,* Holl., and *P. xanthias,* Mab., it forms a small sub-group in the genus.

266. P. XANTHIAS, Mab.   (Plate III. fig. 7.)

Carystus xanthias, Mab. C. R. Soc. Ent. Belg. p. cxvii (1891).
Hab. Lagos *(Mabille);* Gaboon *(Good).*
This species is intermediate between *P. ligora,* Hew., and *P. xanthioides,* Holl.

267. P. XANTHIOIDES, Holl.   (Plate IV. fig. 14.)

Pardaleodes xanthioides, Holl. Ann. & Mag. N. H. (6) vol. x. p. 290 (1892).
Hab. Valley of the Ogové.

268. P. VIBIUS, Hew.

Astictopterus vibius, Hew. Ann. & Mag. N. H. (5) vol. i. p. 343 (1878).
Pamphila rega, Mab. Bull. Soc. Ent. France, (6) vol. ix. p. cxlix (1889); Ann. Soc. Ent. France, (6) vol. x. p. 31, pl. iii. fig. 7, ♀ (1890).
Hab. Tropical West Africa.

269. P. SIERRÆ, sp. nov.   (Plate IV. fig. 19.)

♂. Allied closely to *P. vibius,* Hew. Instead, however, of having the reddish-orange spot on the primaries defined on the lower margin by vein 1, as in that species, this spot extends to the inner margin and likewise inwardly toward the base, being interrupted at the base by a linear patch of raised scales, extending along the lower edge of the cell at the origin of vein 2. The secondaries also are paler on the upper surface, and are marked beyond the cell by an obscure series of yellowish spots. On the underside the wings are much paler than in *vibius,* the secondaries of which on the underside are uniformly black; in this species they are ochraceous, clouded with fuscous and defined externally by a fine black marginal line. This may be a local form of *vibius,* but is sufficiently distinct to deserve description. Expanse 25 mm.

Type in coll. Staudinger.
Hab. Sierra Leone.

270. P. FAN, Holl.

Osmodes (?) *fan*, Holl. Ent. News, vol. v. p. 91, pl. iii. fig. 8 (1894).

Hab. Interior of Cameroons.

After a very careful microscopical study of the anatomical details of the structure of the three preceding species, I can find nothing of generic value to lead me to separate them from the species included in the genus *Pardaleodes*, though in general appearance they present widely different features. The total absence of translucent spots on the primaries, the broader and more rounded character of the wings, and the general style of the markings at first sight appear to reveal such a difference as to have led me for some time to have been inclined to refer these species to the genus *Koruthaialos*, Wats., but the palpi, the neuration, and the antennæ are so exactly in agreement with those of the genus *Pardaleodes*, that I am constrained, in spite of the facies, to place them in the latter genus.

CERATRICHIA, Butl.

271. C. NOTHUS, Fabr.

Papilio nothus, Fabr. Mant. Ins. ii. p. 88 (1787).
Ceratrichia nothus, Butl. Cat. Fabr. Diurn. Lep. pl. iii. fig. 15 (1870); Watson, P. Z. S. 1893, p. 117.

Hab. Tropical West Africa.

This species is not nearly so common as the two following.

272. C. PHOCION, Fabr.

Papilio phocion, Fabr. Spec. Ins. ii. p. 138 (1781).
Ceratrichia phocion, Butl. Cat. Fabr. Diurn. Lep. pl. iii. fig. 14 (1870).
Cyclopides phocæus, Westw., Doubl. & Hew. Gen. Diurn. Lep. p. 251 (1852).
Ceratrichia semilutea, Mabille, C. R. Soc. Ent. Belg. 1891, p. lxv.

Hab. Tropical West Africa.

This species appears to be very common on the Ogové. The female has the primaries profusely spotted in some specimens, and the secondaries more or less suffused with brown, almost obscuring the broad yellow middle area. *Ceratrichia semilutea*, Mab., the type of which is before me as I write, is a slightly dwarfed specimen of the male. Another male, in the Staudinger collection, has been designated as the type of an unpublished species by Mons. Mabille, to which he gives the MS. name *C. punctata*. It is a male with the primaries more spotted than is quite usual, though in a long series of specimens, such as I possess, numerous examples of this form are sure to be found.

273. C. FLAVA, Hew. (Plate III. fig. 14.)

♂. *Ceratrichia flava*, Hew. Ann. & Mag. Nat. Hist. (5) vol. i. p. 343 (1878).

Plastingia charita, Ploetz, S. E. Z. vol. xl. p. 356 (1879).
♀. *Apaustus argyrosticta*, Ploetz, S. E. Z. vol. xl. p. 358
(1879); id. ibid. vol. xlv. p. 156 (1884); id. l. c. A. *argyrospila*,
Ploetz, MS.
Ceratrichia argyrosticta, Wats. P. Z. S. 1893, p. 117.
Hab. Cameroons, Valley of the Ogové, Aburi (*Ploetz*).

This is a very common species. I have an enormous series,
taken at different times and places. There is not a shadow of
doubt in my mind that the above synonymy is correct. The
females are very variable upon the upperside of the wings, but
agree very well with the males on the underside, though in both
the male and the female sex the silvery centres of the spots on the
underside are often suffused with dark brown, and the silvery
colour is rendered obsolete.

COBALUS, Hübn.

274. C. (?) CORVINUS, Mab.

Cobalus corvinus, Mab. Bull. Soc. Ent. France, (6) vol. ix. p. clxix
(1889).
Hab. Sierra Leone (*Mabille*).

I allow this species to remain in this genus, to which it has been
assigned by its author, though it is quite plain to me that it does
not really belong here. I have the type before me as I write, but
as it is unique and in poor condition, so that I cannot without
great risk of further injury make a close anatomical investigation,
I must leave its location undecided. It seems superficially to show
a general relationship to *pulvina*, Ploetz, and *nox*, Mab., but the
wings are more fragile and relatively longer, and the insect is not
so robust.

275. C. (?) ATRIO, Mab.

Cobalus atrio, Mab. C. R. Soc. Ent. Belg. 1891, p. lxxxii.

A figure of the type kindly lent me by Mons. Mabille suggests
that this is the female of *Semalea pulvina*, Ploetz (*q. v.*). It is
certainly not a *Cobalus*, as that genus is South-American.

ANDRONYMUS, gen. nov.

Antennæ more than half the length of the primaries, slender;
club moderate, fusiform, slightly recurved at the tip. Palpi
divergent, with the first and second joints heavily clothed with
scales, the third joint naked, aciculate, erect, as high as the vertex
of the head. Fore wing elongated, with the inner margin con-
siderably longer than the outer margin, blunt at the apex, and
slightly excavated between the extremities of veins 1 and 3. Cell
narrow, elongated, nearly two-thirds the length of the costa; vein
12 reaching the costa before the end of the cell; upper discocellular
short, but distinct, at right angles to the upper margin of cell;
middle discocellular relatively long, curved inwardly; lower disco-

cellular short, forming an obtuse angle with the lower margin of
the cell; lower margin of cell slightly bent outwardly at origin of
vein 2, which is located near the middle of the cell. Vein 3 nearer
to vein 4 than to vein 2. Hind wing with costal margin nearly
straight; outer margin and inner margin rounded; outer and anal
angles broadly rounded. On the upperside of the wing, on the
middle of the fold between veins 7 and 8 near the origin of vein 7,
is a small pencil of long hairs, and vein 6 just beyond the end of

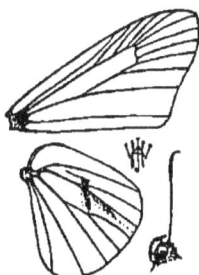

Neuration of *Andronymus philander*, Ploetz. ¼.

the cell is clothed on the underside with a closely appressed bunch
of thick hair-like scales. Discocellulars and vein 5 very faint, if
not quite obsolete. The wings are marked with translucent spots,
those on the primaries being located in the usual order, those on
the secondaries being four in number—a large one at the end of
the cell, and three just below it, one between the origins of veins
2 and 3, and one on either side of this spot, separated from the
central spot by veins 2 and 3. Hind tibiæ with two pairs of
spurs.

Type *A. philander*, Ploetz.

276. A. PHILANDER, Hopff.

Pamphila philander, Hopff. Monatsb. Akad. Wiss. Berl. 1855,
p. 643; Peters' Reise n. Mossamb., Zool. v. p. 416, pl. xxvii.
figs. 1, 2 (1862).
Carystus philander, Kirby, Syn. Cat. p. 590 (1871).
Acleros philander, Butl. P. Z. S. 1893, p. 669.
Ancyloxypha philander, Trim. P. Z. S. 1894, p. 78.
Carystus evander, Mab. Ann. Soc. Ent. France, (6) vol. x. p. 30,
pl. iii. fig. 4 (1890).
Hab. Tropical Western and Central Africa.

277. A. LEANDER, Ploetz.

Apaustus leander, Ploetz, S. E. Z. vol. xl. p. 360 (1879).
Hab. Tropical Western Africa.
This species may be readily distinguished from *A. leander*, its

very near ally, by the yellow colour of the light markings upon the wings.

278. A. NEANDER, Ploetz. (Plate II. fig. 23.)

Apaustus neander, Ploetz, S. E. Z. vol. xlv. p. 154 (1884).
Ancyloxypha producta, Trim. S. Afr. Butt. vol. iii. p. 334 (1889).
Hab. Tropical West Africa; Delagoa.

I have a long series of this species, concerning which Mr. Good wrote me that at the time of capture they appeared to be engaged in migrating in vast numbers. Only upon the occasion of this migration did he observe them during a residence of eight years upon the banks of the Ogové River. Mr. Trimen confirms, after examining specimens I sent him, the opinion I had before communicated to him, that this species is the one named *producta* by him.

HIDARI, Dist.

279. H. CÆNIRA, Hew. (Plate II. fig. 3.)

Hesperia cænira, Hew. Exot. Butt. vol. iv. *Hesperia*, pl. ii. figs. 15, 16 (1867).
Pamphila cænira, Kirby, Syn. Cat. p. 606 (1871).
♀. *Hesperia calpis*, Ploetz, S. E. Z. vol. xl. p. 354 (1879), vol. xliii. p. 328 (1882).
Hab. Gaboon, Cameroons.

The description given by Ploetz of his species named *Hesperia calpis* is unmistakable, if care be taken to make the comparisons which he suggests. I have also been able to identify his species by means of a copy of the figure contained in his plates. For many years I have kept *H. calpis* apart from the older species named *Hesperia cænira* by Hewitson, but upon examination I discover that every specimen of *H. calpis* in my collection, several dozens of them, are females, and all of the specimens of typical *H. cænira* are males. Furthermore, there is such absolute agreement in the markings and coloration of the primaries on the underside of the two forms, as to convince me that they are sexes, and I have accordingly united them as above. The female varies in some instances. I have one specimen in which there is a manifest tendency to an enlargement of the white spot on the primaries, so that the marking approximates more nearly that of the male than is usual.

280. H. LATERCULUS, Holl. (Plate I. fig. 15.)

Proteides laterculus, Holl. Ent. News, vol. i. p. 156 (1890).
Hab. Valley of the Ogové.

281. H. IRICOLOR, Holl. (Plate I. fig. 5.)

Proteides iricolor, Holl. Ent. News, vol. i. p. 156 (1890).
Hab. Valley of the Ogové.

PTEROTEINON, Wats.

(*Tanyptera*, Mab.)

282. P. LAUFELLA, Hew.

Hesperia laufella, Hew. Exot. Butt. vol. iv. *Hesp.* pl. ii. figs. 28–30 (1867).

Carystus laufella, Kirby, Syn. Cat. p. 591 (1871); Staudgr. Exot. Schmett. pl. 99 (1888).

Tanyptera laufella, Mab. Bull. Soc. Zool. France, p. 260 (1877).

Pteroteinon laufella, Wats. P. Z. S. 1893, p. 124.

Hab. Tropical West Africa.

CHORISTONEURA, Mab.

283. C. APICALIS, Mab. (Plate V. fig. 1.)

Choristoneura apicalis, Mab. Bull. Soc. Ent. France, (6) vol. ix. p. clvi (1889).

Hab. Sierra Leone (*Mabille*).

This very remarkable insect is entirely unlike any other species which I have ever seen from the African continent, and recalls in general appearance some of the species of the S. American genus *Entheus*. At the time Lieut. Watson prepared his Revision of the

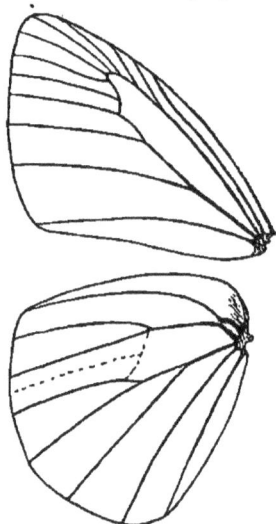

Neuration of *Choristoneura apicalis*, Mab. ⅞.

genera of the Hesperiidæ, no specimen of this insect was available by him for purposes of study. I take pleasure in incorporating a cut giving the neuration. From this it will be seen by the student

6*

that the neuration is quite peculiar, and that Mons. Mabille, the author of the genus, was abundantly justified by the facts in erecting it for the reception of the typical species.

GAMIA, gen. nov.

Antennæ long, slender; club robust, tapering gradually, produced at the apical extremity to a fine point, which is slightly recurved.   Palpi: first joint short; second joint long, both heavily clothed with hair; the third joint long, produced and conical, almost naked; the hind tibiæ with a double pair of spurs, and heavily clothed with long hair. Fore wing: inner margin longer than outer margin; the costa evenly rounded; the apex obtuse: the outer margin slightly excavated above the outer angle; cell more than two-thirds the length of costa; vein 12 reaches the costa before the end of the cell; vein 5 very slightly nearer vein 4 than vein 6; vein 7 from the end of the cell, very near vein 6; vein 3 very near vein 4, from near the lower angle of the cell;

Antennæ and palpi of *Gamia galua*, Holl. ?.

vein 2 from one-third of the distance from the base to vein 3. Secondaries: costal and outer margins evenly rounded, produced at the anal angle and slightly truncated at anal angle; vein 5 present and distinct; vein 4 from the lower angle of the cell; vein 3 slightly before the lower angle; vein 2 twice as far from vein 3 as the latter is from vein 4; vein 7 from about the middle of the cell.—The insects belonging to this genus are large in size, dark in colour, with the primaries and secondaries ornamented with large translucent yellow spots.   *G. buchholzi* is the largest of all the African Hesperiidæ, with the exception of *Rhopalocampta ithis*.   They are distinctly separate from the genus *Cænides*, to which they are apparently allied by the peculiar form of the palpi.

Type *G. galua*, Holl.

.284. G. GALUA, Holl. (Plate I. fig. 1, ♀.)

Proteides galua, Holl. Ent. News, vol. ii. p. 3 (1801).

Hesperia zintgraffi, Karsch, Ent. Nachr. vol. xviii. p. 178 (1892).

? *Proteides ditissimus*, Mab. C. R. Soc. Ent. Belg. vol. xxxv. p. cxii (1891). ·

Hab. Tropical West Africa.

A comparison of my species with the type of *H. zintgraffi*, Karsch, shows the two to be identical.   I am also strongly inclined to the opinion that *P. ditissimus*, Mab., is the same insect. Unfortunately I have not seen the type of *P. ditissimus*. Mons. Mabille affirmed the identity of the two species when examining my type, but has since expressed in letters a different opinion.

285. G.(?) DITISSIMUS, Mab.

Proteides ditissimus, Mab. C. R. Soc. Ent. Belg. vol. xxxv. p. cxii (1891).

Hab. Sierra Leone (*Mabille*).

Very probably the same as the foregoing species (*q. v.*).

286. G. BUCHHOLZI, Ploetz.

♀. *Hesperia buchholzi*, Ploetz, S. E. Z. vol. xl. p. 354 (1879), vol. xliii. p. 330 (1882).

Gangara (?) *basistriga*, Holl. Ent. News, vol. v. p. 29, pl. i. fig. 12 (1894).

Hab. Aburi (*Ploetz*); Ogové (*Holland*).

Strangely enough, none but females of this species have been found thus far. The type was a unique female in the collection made by Buchholz. There is another specimen in my collection, and another still in the hands of Mons. Mabille, to which he has affixed the MS. name "*robustus*."

CÆNIDES, gen. nov.

Antennæ long, slender; club moderate, long, produced at the apical extremity to a long fine point, bent back at a right angle. Palpi: first joint short, second joint long, erect, reaching the tip of the vertex, both densely clothed with long hair; third joint minute, erect, and almost concealed by the hairy vestiture of the second joint. Primaries with the inner margin longer than the outer margin, or, in some species, subequal. Cell slightly less than two-thirds the length of the costa; vein 12 of the primaries terminating before the end of the cell; vein 7 arising slightly before the end of the cell; vein 5 much nearer 4 than 6; vein 3 near vein 4; vein 2 from about the middle of the lower margin of the cell. The secondaries with vein 5 obsolete, or very faintly visible; discocellulars faint, angulated, with the point of the angle turned toward the base; cell short. Legs armed with double sets of spurs on the hind tibiæ.

The species of this genus, which is a large one, may be arranged in four groups. The first is represented typically by *C. dacela*, Hew., in which the primaries of the male have a sexual curved stigma below the cell crossing veins 3 and 2, and a large oval patch of raised, glossy hairs upon the outer end of the cell of the secondaries, covering the origin of veins 2, 3, and 4, and extending beyond toward the outer margin. The second group is represented by species in which the large oval patch of raised scales on the secondaries is absent, or at most represented by a tuft of loose and not conspicuous hairs. The discal band of the primaries is present. This group is composed of species of which *C. maracanda* and *C. leonora* are typical. The third group is composed of species in which the sexual brand of the primaries in the male is absent, while the large oval patch of hairs in the secondaries remains.

This division of the genus contains *C. benga* and possibly others.
The fourth group is composed of species in which both secondaries
and primaries are without sexual brands or marks of a conspicuous
and easily discernible character, the brands and patches of raised
scales being revealed in some of the forms only after bleaching and
microscopic examination, and then as merely obsolescent.

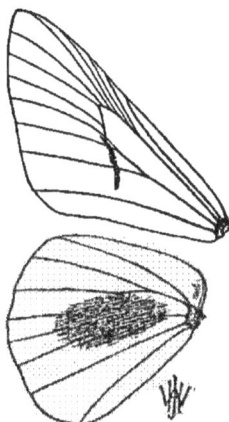

Neuration of *Cænides dacela,* Hew. ♂.

In the species of all these four groups the antennæ, the palpi,
the neuration, and the outline of the wings are the same. They
are differentiated into groups by the sexual markings of the male
sex, so far as my studies have informed me. Most of the species
have been hitherto referred by writers to the genus *Proteides*, to
which they manifestly do not belong.

287. C. DACELA, Hew. (Plate II. fig. 2, ♂ ; Plate V. fig. 18, ♀.)

Hesperia dacela, Hew. Ann. & Mag. Nat. Hist. (4) vol. xviii.
p. 451 (1876).

Hesperia nydia, Ploetz, S. E. Z. vol. xl. p. 353 (1879), vol. xliii.
p. 326 (1882).

♀. *Plastingia podora*, Ploetz, S. E. Z. vol. xlv. p. 150 (1884).

As to the identification of the male of this species with the insect
described by Ploetz as *Hesperia nydia*, there is not a shadow of
doubt in my mind. The insect described by Ploetz as *Plastingia
podora* was contained in the Berlin Museum. The insect labelled
as such was examined for me both by Dr. Karsch and Dr. Scudder,
and is represented in the plates accompanying this article, being
reproduced after a careful drawing by Von Prillwitz. It is unmistakably the female of *C. dacela*. Unfortunately, however, the
description given by Ploetz of his *P. podora* does not tally with the

insect which, bearing his own manuscript label, is accepted as the type. I have been puzzled to find a satisfactory solution of the difficulty, but have resolved to accept the authenticated type specimen as the key to the problem, and have therefore given the synonymy as above. Of course it is quite possible that a misplacement of the original label may have taken place, but at this distance, both of space and time, I am not in a position to clear up the difficulty. The description given by Ploetz is, as usual, not clear enough to help to a positive conclusion as to what he meant by it.

288. C. SORITIA, How. (Plate I. fig. 9.)

♂. *Hesperia soritia*, Hew. Ann. & Mag. Nat. Hist. (4) vol. xviii. p. 453 (1876).

♀. *Proteides xychus*, Mab. C.R. Soc. Ent. Belg. vol. xxxv. p. cxi (1891).

Proteides xantho, Mab. C.R. Soc. Ent. Belg. vol. xxxv. p. cxi (1891).

Hab. Gaboon, Sierra Leone.

Upon a comparison of the types of *P. xychus* and *P. xantho*, Mab., with the type of *H. soritia*, Hew., it becomes plain that they are one and the same species. The females vary in the amount of maculation on both the upper and under side of the secondaries. Some specimens have a distinct pale discal spot at the end of the cell upon the lower side of the secondaries, followed by a discal curved series of similar small spots, frequently obscurely visible upon the upper surface; other specimens are almost devoid of these markings, which are generally more or less obsolescent. A female with these markings more distinct than usual was selected by Mons. Mabille as the type of his *xantho*. It is before me as I write, and I cannot feel justified in regarding it as separate from *C. soritia*. In a long series of specimens of *soritia*, such females are not at all uncommon.

289. C. KANGVENSIS, sp. nov. (Plate I. fig. 10.)

♂. Body with palpi and antennæ, as well as legs, brown, the under surfaces slightly paler than the upper surfaces. The wings are brown, somewhat inclining to tawny fuscous at the base. The cilia are pale fuscous. The primaries are marked with three minute subapical spots, arranged in a curved series, by a large quadrate spot at the end of the cell, which is notched on its outer margin, and by two moderately large subquadrate spots, lying one on either side of vein 3 at its origin, the lower spot being the largest. There is a fine raphe, or sexual brand, running along the inner margin of this large spot and continued across interval 1 toward the inner margin. The secondaries have the end of the cell and a portion of the disc immediately beyond the end covered by a large oval patch of raised glossy black hairs. On the underside the primaries are paler on the apical third, with the inner margin broadly pale testaceous. The translucent spots of the upper surface reappear on this side, though less distinctly defined, owing to the paler ground-colour. The secondaries are dark brown, slightly touched with greyish on the outer margin near the outer angle. There are a few obscure palo discal spots beyond the cell.

♀. The female is marked like the male, but lacks, of course, the characteristic sexual markings of the male. The wings are more elongated and rounded, and the primaries have a translucent yellow spot on interval 1, midway between the base and the outer margin.

Expanse, ♂ 40 mm., ♀ 43 mm.

Hab. Valley of the Ogové.

This species is closely related to *C. soritia,* Hew., but is quite distinct.

290. C. MARACANDA, Hew. (Plate I. fig. 4.)

Hesperia maracanda, Hew. Ann. & Mag. Nat. Hist. (4) vol. xviii. p. 450 (1876).

Casyapa masacanda, Kirby, Syn. Cat. Suppl. p. 817 (1877).

Hab. Angola (*Hewitson*); Gaboon (*Good*).

291. C. BINOEVATUS, Mab. (Plate II. fig. 1.)

Proteides binoevatus, Mab. C. R. Soc. Ent. Belg. vol. xxxv. p. cxii (1891).

Hab. Valley of the Ogové.

292. C. LEONORA, Ploetz. (Plate II. fig. 5.)

Hesperia leonora, Ploetz, S. E. Z. vol. xl. p. 355 (1879), vol. xliii. p. 338 (1882).

Proteides xanthargyra, Mab. C. R. Soc. Ent. Belg. vol. xxxv. p. cxii (1891).

Pamphila leonora, Karsch, Berl. Ent. Zeit. vol. xxxviii. p. 253 (1893).

Hab. Aburi (*Ploetz*); Accra (*Mabille*); Togoland (*Karsch*); Valley of Ogové (*Good*).

The number of the small subapical spots in this species is variable. Some specimens have but two, others three, while the type of Ploetz is destitute of such spots. The absence of the sexual brand on the upperside of the primaries of the male is apparently the only mark of distinction having generic weight which would lead me to separate this species from the foregoing three. If there are other points, I have failed to discover them, and I hesitate to erect a new genus for the reception of this species without some more evident reason.

293. C. STOEHRI, Karsch.

Pamphila stoehri, Karsch, Berl. Ent. Zeit. vol. xxxviii. p. 252, pl. vi. fig. 6 (1893).

Hab. Togoland (*Karsch*); Gaboon (*Mocquerys*).

The type was a damaged male. The collection of Dr. Staudinger contains two perfect females of this fine species, taken at Gaboon by Mocquerys according to the labels. The female is like the male, but larger in size, and with the underside of the wings redder than in the figure of the type given by Karsch. It is singular that during the eight years in which I have had a collector constantly residing and at work for me in French Congo, this species has not turned up. It evidently must be very rare, or very local in its distribution.

294. C. BENGA, Holl. (Plate I. fig. 13.)

Proteides benga, Holl. Ent. News, vol. ii. p. 4 (1891).

Hab. Valley of the Ogové.

295. C. CYLINDA, Hew. (Plate I. fig. 12.)

Hesperia cylinda, Hew. Ann. & Mag. Nat. Hist. (4) vol. xviii. p. 449 (176).

Pamphila calpis, Karsch (*nec* Ploetz), Berl. Ent. Zeit. vol. xxxviii. p. 252, pl. vi. fig. 4 (1893).

(*Proteides ruralis*, Mab. MS., *cf.* Staudinger's price-lists.)

Hab. Tropical Western Africa. Very common at Gaboon.

This species has been labelled *P. ruralis* by Mons. Mabille in several collections, and has been sold under this name by Dr. Staudinger. I can find no account of the publication of the species by Mons. Mabille, and believe the name to be hitherto unpublished, except as stated, and as it is once or twice referred to in the writings of Mons. Mabille. It seems at all events to have totally escaped the notice of the compilers of the 'Zoological Record' and Bertkau's 'Register,' and, though I have twice asked Mons. Mabille to inform me where the species is described, he has failed to include an answer to this question with the other information he has so kindly and generously given me. The identification of this species with *P. calpis*, Ploetz, by Dr. Karsch is based upon specimens so labelled in the Berlin Museum; but these are not types, and came from Senegal, and were not labelled by Ploetz. There is, further, no agreement whatever between the insect figured by Karsch and the description of *P. calpis* given by Ploetz. A comparison of the figure given by Karsch shows the entire identity of the insect with Hewitson's *H. cylinda*. The true *calpis* is figured in this paper. It is the female of *Hidari cœnira*, Hew.

C. *cylinda* is a crepuscular insect, as I have been informed by the late Dr. Good. It only appears at dusk in the morning or the evening, though occasionally on dark and cloudy days it may be seen upon the wing. I have one or two examples which were taken at lamp-light, having flown into the room after dark.

296. C. DACENA, Hew.

Hesperia dacena, Hew. Ann. & Mag. Nat. Hist. (4) vol. xviii. p. 453 (1876).

Proteides leucopogon, Mab. C. R. Soc. Ent. Belg. vol. xxxv. p. cxi
(1891); Nov. Lepidopt. p. 111, pl. xv. fig. 5 (1893).

Hab. Gaboon, Cameroons.

297. C. ORMA, Ploetz.

Ismene orma, Ploetz, S. E. Z. vol. xl. p. 363 (1879), vol. xlv.
p. 59 (1884).
Hesperia violascens, Ploetz, S. E. Z. vol. xliii. p. 322 (1882).
Pamphila violascens, Karsch, Berl. Ent. Zeit. vol. xxxviii. p. 248,
pl. vi. fig. 3 (1893).

Hab. Cameroons, Ogové Valley.

H. violascens was described, as Dr. Karsch has shown, from a
drawing of the upperside of a specimen named *violascens* by
Maassen. Had Ploetz seen the specimen from which the drawing
was made, he would no doubt have recognized in it his own
I. orma. The underside is unmistakable. Dr. Karsch has
correctly determined the species as *violascens*, Ploetz, but has
failed to recognize its identity with the species described as *orma*
by Ploetz, and referred by him to the genus *Ismene.* This
reference is sufficiently exact to suffice, though I have been
inclined to create a subdivision of the genus for the reception of
this species, owing to the fact that the antennæ are not so greatly
swollen below the tip as in the other species of the genus, and the
outer margin of the secondaries is not so strongly excavated before
the anal angle. It is worthy of note that the white band on the
underside of the secondaries varies greatly, and in some specimens
is reduced to a narrow line, and in others is almost obsolete.

298. C. CORDUBA, Hew.

Hesperia corduba, Hew. Ann. & Mag. Nat. Hist. (4) vol. xviii.
p. 454 (1876).
Proteides massiva, Mab. & Vuill. Nov. Lepidopt. p. 21, pl. iii.
fig. 4 (1891).

Hab. Gaboon, Sierra Leone.

This species is very common in the Valley of the Ogové. Thus
far, singularly enough, I have never seen a male specimen. Of
the twenty-five, or more, examples in my collection, all appear to
be females.

299. C. WAGA, Ploetz.

Telesto waga, Ploetz, S. E. Z. vol. xlvii. p. 108 (1886).

Hab. Aburi (*Ploetz*).

From a copy of the figure of this species contained in the unpublished collection of drawings made by Herr Ploetz, and to
which he refers in his descriptions, this insect is closely allied to
C. cylinda, Hew., and, if I am not greatly mistaken, the drawing
represents a rubbed specimen of *C. cylinda*; certainly specimens of
cylinda in poor condition agree extremely well with the figure of
Ploetz.

300. C. ILERDA, Moeschler.

Hesperia ilerda, Moeschler, Abhandl. Senckenb. naturf. Ges. Bd. xv. p. 65, pl. i. fig. 16 (1887).

Pamphila ilerda, Karsch, Berl. Ent. Zeit. vol. xxxviii. p. 251 (1893).

Hab. Tropical West Africa.

I have specimens of what are undoubtedly *C. cylinda*, Hew., which agree absolutely with the figure of *ilerda* given by Moeschler. Unfortunately Moeschler does not give a representation of the underside of his specimen, and I am therefore left in doubt as to whether the two species are identical.

301. C. LACIDA, Hew.   (Plate I. fig. 14.)

Hesperia lacida, Hew. Ann. & Mag. Nat. Hist. (4) vol. xviii. p. 453 (1876).

Hab. Gaboon (*Hewitson*).

The type of Hewitson is a female.

302. C. ZAREMBA, Ploetz.   (Plate V. fig. 5.)

Telesto zaremba, Ploetz, S. E. Z. vol. xlv. p. 376 (1884).

Hab. Old Calabar (*Ploetz*); French Congo (*Mocquerys*).

There are two somewhat damaged specimens in the collection of Dr. Staudinger. The reference to this genus seems proper, though, in the rubbed condition of the upperside of the secondaries of both examples, I am unable to make sure of the presence of the tuft of long hairs upon the cell which is characteristic of most of the species of the genus.

303. C. BALENGE, Holl.   (Plate I. fig. 3.)

Proteides balenge, Holl. Ent. News, vol. ii. p. 4 (1891).

Hab. Valley of the Ogové.

The type is a female, and remains so far unique in my collection. A fine male is contained in the collection of Dr. Staudinger. These are, so far as I know, the only examples extant in the museums of the world of this fine species, which is one of the largest of the African Hesperiidæ. The female and the male do not differ materially, except in size and the form of the wings, as is usual.

304. C. SEXTILIS, Ploetz.

Hesperia sextilis, Ploetz, S. E. Z. vol. xlvii. p. 89 (1886); Moeschler, Abhandl. Senck. naturf. Ges. Bd. xv. p. 64 (1887).

Hab. Aburi (*Ploetz*).

This species is stated by Moeschler to belong to the same group as *C. calpis*, Ploetz, by which sign it might be located in the genus *Hidari*, were it not for the fact that in some way or other some German authors have come to traditionally regard the insect named *cylinda* by Hewitson as being the one designated as *calpis* by Ploetz. Moeschler is one of the authors who held this view, and hence I place *sertilis* in the same group as *cylinda*. I do not know the species under this name at all events.

305. C. (?) PROXIMA, Ploetz.

Hesperia proxima, Ploetz, S. E. Z. vol. xlvii. p. 95 (1886).

Hab. West Africa (*Ploetz*).

I only know this species from a copy of the drawing by Ploetz. In the form of the wings it suggests affinity to the species which I have located in the genus *Cænides*, but it probably does not belong there.

ARTITROPA, gen. nov.

Antennæ moderately long, more than half the length of the costa of the primaries; club robust, elongated, terminating in a short fine point slightly recurved. Palpi stout, erect, reaching the top of the vertex; the second and third joints are densely clothed with hair; the third joint is minute, almost concealed in the vestiture of the second joint. The legs have the tibiæ scantily

Neuration of *Artitropa erinnys*, Trim., ♂. ¼.

clothed with long hair; those of the posterior pair are armed with a median and double terminal spurs. The primaries have the costa slightly rounded; the inner and the outer margins are subequal, evenly rounded; the cell is two-thirds the length of the costa, with the upper angle acute, the lower angle obtuse; vein 5 slightly nearer vein 4 than vein 6; vein 12 terminates on the costa before the end of the cell; vein 7 arises slightly before the end of the cell; vein 2 is more than twice as far from vein 3 as vein 3 is from vein 4 and is equidistant between vein 3 and the base. The cell of the secondaries is short; vein 5 is present and distinct; vein 3 and vein 7 arise well before the end of the cell; the outer margin is rounded and slightly excavated above the termination of vein 1*b*.

Type *A. erinnys*, Trimen.

I have erected this genus for the reception of the following species, which are distinguished from all other near allies in the genus *Cænides* and allied genera by the shape of the club of the antennæ, by their more robust form, and by their peculiar style of coloration. They form a well-marked group.

306. A. ERINNYS, Trim.

Pamphila erinnys, Trim. Trans. Ent. Soc. Lond. (3) vol. i.
p. 290 (1861); Rhop. Afr. Austr. vol. ii. p. 303, pl. vi. fig. 8
(1866); S. Afr. Butt. vol. iii. p. 326 (1889).

Hab. Southern Africa.

307. A. COMUS, Cram.

Papilio comus, Cram. Pap. Exot. iv. pl. 391. figs. N, O (1782).
Papilio helops, Dru. Ill. Ex. Ent. iii. pl. xxxiii. figs. 2, 3 (1782).
Hesperia ennius, Fabr. Ent. Syst. iii. 1, p. 337 (1793); Latr. Enc.
Méth. vol. ix. p. 749 (1823).
Papilio ennius, Don. Ins. India, p. 59, pl. li. fig. 1 (1800).
Proteides helops, Butl. Cat. Fabr. Diurn. Lep. p. 265 (1869);
Kirby, Syn. Cat. p. 595 (1871).
Pamphila comus, Karsch, Berl. Ent. Zeit. vol. xxxviii. p. 249
(1893).

Hab. West Africa.   (Err. "*Surinam*," Cram.; "*In Indiis*,"
Drury.)

308. A. MARGARITATA, Holl.   (Plate I. fig. 2.)

Proteides margaritata, Holl. Ent. News, vol. i. p. 155 (1890).

Hab. Valley of the Ogové.

I have been inclined to regard this species as identical with
A. comus, Cram. But an examination of specimens made for me
by my good friend Dr. S. H. Scudder, at Berlin and at the British
Museum, he having in his possession at the time the drawing which
is reproduced in the Plate, casts a great doubt upon the correctness
of this view. Dr. Scudder says, " Your *margaritata* is most certainly not the insect labelled *helops=comus* in the British Museum,
and is very doubtfully the insect known as *comus*, in the Museum
in Berlin." I had sunk my name as a synonym until receiving
this opinion from my learned friend, who is recognized as a very
high authority in all such matters.

309. A. BOSEÆ, Saalm.

Hesperia boseæ, Saalm, Lep. von Madgr. p. 105, pl. i. figs. 15, 16
(1884).
Proteices boseæ, Mab. Grand. Madgr. vol. xviii. p. 329, pl. lii.
figs. 10, 10 a (1887).

Hab. Madagascar.

309 a. A. SHELLEYI, Sharpe [1].

Proteides shelleyi, E. M. Sharpe, Ann. & Mag. N. H. (6) vol. vi.
p. 349 (1890).

Hab. Fantee (*Capt. Shelley*).

[1] Unfortunately this species was by an oversight omitted when the MS. was
in preparation.

PLOETZIA, Saalm.

(*Systole*, Mab.)

310. P. AMYGDALIS, Mab.

Hesperia amygdalis, Mab. Bull. Soc. Zool. France, 1877, p. 234.
Systole amygdalis, Mab., Grandidier's Madagascar, vol. xviii.
p. 330, pl. li. figs. 6, 6 *a*, 7 (1887).
Ploetzia amygdalis, Saalm. Lep. Madgr. vol. i. p. 115 (1884).
Hab. Madagascar.

311. P. FIARA, Butl.

Proteides fiara, Butl. Trans. Ent. Soc. Lond. 1870, p. 503;
Staudgr. Exot. Schmett. vol. ii. pl. 99 (1888).
Pamphila fiara, Trim. S. Afr. Butt. vol. iii. p. 329 (1889).
Hab. South Africa.

312. P. DYSMEPHILA, Trim.

Pamphila dysmephila, Trim. Trans. Ent. Soc. Lond. 1868, p. 96,
pl. vi. fig. 10.
Hesperia mucorea, Karsch, Ent. Nachr. vol. xviii. p. 178 (1892).
Hab. South Africa, Togoland.
Through the kindness of Dr. Karsch I have been permitted to
have a carefully drawn figure of his *Hesperia mucorea* executed by
Herr Prillwitz, and it proves upon comparison with typical specimens of the male of *P. dysmephila*, received from Mr. Trimen, to
be the same. The absence of the white line upon the underside
of the secondaries, which is so conspicuous in the female, and is
brought out characteristically in the figure given by Mr. Trimen, is
calculated to mislead the student who is not aware of this difference in the marking of the sexes.

313. P. CERYMICA, Hew.

Hesperia cerymica, Hew. Ex. Butt. iv. *Hesp.* pl. ii. figs. 20,
21 (1867).
Carystus cerymica, Kirby, Syn. Cat. p. 591 (1871); Trim. S. Afr.
Butt. vol. iii. p. 329, footnote (1889).
Hab. Tropical West Africa.
Mr. Trimen is quite right in his surmise expressed on p. 329 of
the third vol. of his 'S. African Butterflies.'

314. P. QUATERNATA, Mab.

Pamphila quaternata, Mab. Bull. & Ann. Soc. Ent. France, (5)
vol. vi. pp. xxvi & 268 (1876).
Hab. Senegal (*Mabille*).
This species is stated by the author to be very closely allied to
P. dysmephila, Trim. The type was unique.

315. P. CAPRONNIERI, Ploetz.

Hesperia capronnieri, Ploetz, S. E. Z. vol. xl. p. 353 (1879),
vol. xliii. p. 326 (1882).
Proteides capronnieri, Mab. Ann. Soc. Ent. France, (6) vol. x.
p. 33, pl. iii. fig. 3 (1890).
Hab. Aburi (*Ploetz*), Cameroons (*Mabille*).

This is a very distinct species. The female lacks the broad
white anterior margin on the upperside of the costal area of the
secondaries which is so conspicuous a feature in the male.

316. P. WEIGLEI, Ploetz.

Hesperia weiglei, Ploetz, S. E. Z. vol. xlvii. p. 90 (1886);
Moeschler, Abhandl. Senck. naturf. Ges., Bd. xv. p. 65, pl. i. fig. 18
(1887).
Pamphila weiglei, Karsch, Berl. Ent. Zeit. vol. xxxviii. p. 253
(1893).
Hab. Tropical West Africa.

I am strongly inclined to think that this species is only a form
of *P. cerymica*, Hew.

317. P. NOBILIOR, sp. nov. (Plate V. fig. 2.)

♀. The antennæ are marked with white on the lower side of
the club. The body above and below and the wings upon the
upperside are tawny fuscous. The primaries are marked by four
waxen yellow translucent spots in the cell near its end, and by two
similar discal spots, one on either side of vein 3 near its origin.
Of the two spots in the cell the upper one is very small and the
lower is much larger, oval, produced. The two discal spots are
subquadrate, and the lower one is thrice the size of the upper one.
The cilia are paler than the body of the wing, and the costa is also
paler toward the base. On the underside both wings are rich
dark maroon, growing paler towards the outer margin. The
nervules are more or less white and stand out distinctly upon the
darker ground, especially at their extremities on the primaries, and
in the case of veins 6, 7, and 8 on the secondaries. The triangular
space on the secondaries between veins 6 and 7 is perceptibly paler
than the rest of the wing. The translucent spots appear upon the
lower surface of the primaries as upon the upperside, and in
addition the inner margin of the primaries is pale testaceous. The
secondaries have a minute white spot in the cell near its end, and
two similar white spots, one on either side of vein 2 about midway between its origin and the outer margin.
Expanse 48 mm.
Hab. Lambarene, French Congo (*Mocquerys*).
The type is in the collection of Dr. Staudinger.

ACALLOPISTES, gen. nov.

Antennæ slender, more than half as long as the costa of the
primaries; club about one fourth the length of the entire antennæ,
suddenly enlarging and then gradually tapering to the tip, gently

recurved. The palpi are short, with the first and second joints densely clothed with hairs, the third joint minute and almost concealed by the vestiture of the second joint. The tibiæ are clothed with long hairs, and those of the hind legs are armed with

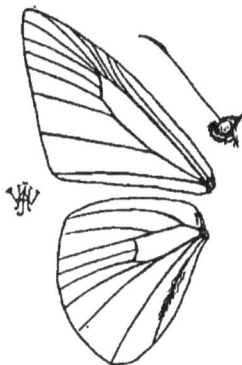

Head and neuration of *Acallopistes holocausta*, Mab., ♂. ♀.

double terminal spurs. The anterior wings are subtriangular, with the inner and outer margins subequal and straight. The costa is evenly rounded, the apex is acute. The cell of the primaries is a little less than two-thirds the length of the costa, with the upper angle acute and the lower angle obtuse. Vein 12 reaches the costa before the end of the cell ; vein 5 is slightly nearer vein 4 than vein 6 ; veins 6, 7, and 8 rise from about the upper angle of the cell ; vein 3 is twice as far from vein 2 as from vein 4 ; vein 2 is equidistant between the base and vein 3. The secondaries have the costa relatively straight. The outer margin is evenly rounded to the extremity of vein 1 b, at which the wing is produced somewhat sharply. The inner margin is gently rounded and somewhat excavated before the base. The cell is less than half the distance from the base to the outer margin. Vein 5 is distinct. Vein 2 arises beyond the middle of the lower margin of the cell, vein 3 a little before its end. Vein 7 arises from well before the end of the cell, and vein 3 twice as far from vein 7 as from the base.

Type *A. holocausta*, Mab.

The two species referable to this genus are moderately large insects, uniformly dark in colour and without any conspicuous markings.

318. A. HOLOCAUSTA, Mab.   (Plate V. fig. 13.)

Erinota holocausta, Mab. C. R. Soc. Ent. Belg. 1891, p. cxi.

Hab. Cameroons (*Mabille*).

This insect is not an *Erinota*, nor in any way nearly related to the insects properly included in that genus. I find it more closely allied to the insects belonging to that section of the genus *Rhopalocampta* which contains *R. unicolor*, Mab., and *R. libeon*, Druce, but thoroughly separate from them by reason of the different structure of the palpi and the antennæ.

319. **A. DIMIDIA**, sp. nov. (Plate V. fig. 7.)

♂. Antennæ, body, and wings both above and below uniformly dark brown, with a slight greenish sheen on the disc of the primaries when viewed in strong sunlight. The palpi on the lower side are orange-coloured. Expanse 40 mm.

Hab. Gaboon (*Mocquerys*).

The type of this insect is contained in the collection of Dr. Staudinger and is unique. On comparison with *A. holocausta*, Mabille, the chief points of difference are the smaller size and the more obscure colouring, for *A. holocausta* has the primaries and secondaries somewhat plentifully sprinkled with golden-orange scales near the base, and the general coloration is brighter. There is no doubt in my mind as to the specific distinctness of this form upon comparison. The facies is quite distinct, though the species are very closely related.

RHOPALOCAMPTA, Wallgr.

320. **R. RAMANETEK**, Boisd.

Thymele ramanetek, Boisd. Faune Entom. Madgr. p. 62, pl. ix. fig. 3 (1833).

Ismene ramanetek, Kirby, Syn. Cat. p. 581 (1871); Mab., Grandid. Madgr. vol. xviii. p. 326, pl. ll. figs. 2, 2 *a* (1887).

Rhopalocampta ramanetek, Watson, P. Z. S. 1893, p. 129.

Hab. Madagascar.

321. **R. UNICOLOR**, Mab.

Ismene unicolor, Mab. Ann. Soc. Ent. France, (5) vol. vii. p. xxxix (1877); Bull. Soc. Zool. France, 1877, p. 230.

Hesperia unicolor, Trim. S. Afr. Butt. vol. iii. p. 375.

Hab. South Africa, Western Africa as far north as Liberia. Very common on the Ogové River.

322. **R. LIBEON**, Druce.

Ismene libeon, Druce, P. Z. S. 1875, p. 416.

Hesperia libeon, Trim. S. Afr. Butt. vol. iii. p. 375.

Rhopalocampta libeon, Watson, P. Z. S. 1893, p. 129.

Hab. Angola (*Druce*).

Closely allied to *R. unicolor*, Mab.

323. **R. BRUSSAUXI**, Mab.

Ismene brussauxi, Mab. Bull. Soc. Ent. France, 1890, p. ccxxi.

Hab. French Congo (*Mabille*).

This species was described by Mons. Mabille from a defective example. It is evidently very near *R. libeon* and *R. unicolor*.

www.ingramcontent.com/pod-product-compliance
Lightning Source LLC
Chambersburg PA
CBHW021827190326
41518CB00007B/767